Climate Smart Greenhouses - Innovations and Impacts

Edited by Ahmed A. Abdelhafez and Mohamed H.H. Abbas

Published in London, United Kingdom

Climate Smart Greenhouses - Innovations and Impacts
http://dx.doi.org/10.5772/intechopen.107801
Edited by Ahmed A. Abdelhafez and Mohamed H.H. Abbas

Contributors
Taiwo Bintu Ayinde, Charles Fredrick Nicholson, Benjamin Ahmed, Zainab Abdel Mo'ez Mansour Embaby, Geordie Zapalac, Nirmali Bordoloi, Raushan Kumar, Božidar Benko, Sanja Fabek Uher, Sanja Radman, Nevena Opačić, Ahmed A. Abdelhafez, Mohamed H.H. Abbas, Shawky M. Metwally, Hassan H. Abbas, Amera Sh.Metwally, Khaled M. Ibrahim, Aya Sh. Metwally, Rasha R.M. Mansour, Xu Zhang, Oludolapo Akanni Olanrewaju, Oluwafemi Ezekiel Ige, Busola Dorcas Akintayo, Ahad Ali

First published in London, United Kingdom, 2024 by IntechOpen
IntechOpen is the global imprint of INTECHOPEN LIMITED, registered in England and Wales, registration number: 11086078, 5 Princes Gate Court, London, SW7 2QJ, United Kingdom
Printed in Croatia

British Library Cataloguing-in-Publication Data
A catalogue record for this book is available from the British Library

Additional hard and PDF copies can be obtained from orders@intechopen.com

Climate Smart Greenhouses - Innovations and Impacts
Edited by Ahmed A. Abdelhafez and Mohamed H.H. Abbas
p. cm.

This title is part of the Agricultural Sciences Book Series, Volume 4
Topic: Agronomy and Horticulture
Series Editor: W. James Grichar
Topic Editor: Ibrahim Kahramanoglu
Associate Topic Editors: Murat Helvaci and Olga Panfilova

Print ISBN 978-1-83768-975-0
Online ISBN 978-1-83768-976-7
eBook (PDF) ISBN 978-1-83768-977-4
ISSN 3029-052X

IntechOpen Book Series
Agricultural Sciences
Volume 4

Aims and Scope of the Series

The importance of agriculture cannot be overstated. It helps sustain life, as it gives us the food we need to survive and provides opportunities for economic well-being. Agriculture helps people prosper around the world and combines the creativity, imagination, and skill involved in planting crops and raising animals with modern production methods and new technologies. This series includes two main topics: Agronomy and Horticulture, and Animal Farming. This series will help readers better understand the intricacies of production agriculture and provide the new knowledge that is required to be successful. The success of a farmer in modern agriculture requires knowledge of events happening locally as well as globally that impact input decisions and ultimately determine net profit.

Meet the Series Editor

W. James Grichar has been employed with Texas A&M AgriLife Research for over 45 years with an emphasis on research in agronomy, plant pathology, and weed science. He obtained his BS from Texas A&M in 1972 and his Masters of Plant Protection in 1975. He has published 195 journal articles, over 330 research reports and briefs, 11 book chapters, and over 300 abstracts of profession meetings. He also directs research in many crops including corn, grain sorghum, peanuts, and sesame. He has held various positions in different professional societies including the American Peanut Research and Education Society, Southern Weed Science Society, and Texas Plant Protection Conference in addition to being Associate Editor for Peanut Science and Weed Technology. Significant accomplishments have included spearheading efforts to determine the optimum planting time for soybean production along the upper Texas Gulf Coast. These efforts have shown growers that soybean yields can be improved by 10 to 20% by following a late March to early April plant date. He also has been instrumental in developing a herbicide program for peanut production in the south Texas growing region. Through the development and use of herbicides that are effective against major weed problems in the south Texas region, peanut yields have increased by 25 to 30%.

Meet the Volume Editors

Professor Ahmed A. Abdelhafez is the head of the Department of Soils and Water Science at New Valley University, Egypt. Renowned as a leading expert on biochar in the Arab region, he has contributed more than a decade to research at the Agricultural Research Center (ARC). Specializing in agricultural production, environmental contamination control, and biochar technology, Professor Abdelhafez is affiliated with the National Committee of Soil Sciences and the Academy of Scientific Research & Technology. He has published extensively on environmental risks and remediation and serves as the chief editor for *Biochar and Compost Technology*. His innovative research and leadership continue to influence the agricultural and environmental spheres.

Professor Mohamed H.H. Abbas is a distinguished Professor of Soil Chemistry at the Faculty of Agriculture, Benha University, Egypt. Embarking on his academic journey as a demonstrator at Zagazig University, Egypt, in 2000, Professor Abbas earned both his MA and Ph.D. from Benha University. A dedicated board member of the *Egyptian Journal of Soil Science*, his research intricately weaves together themes such as the sorption/desorption dynamics of potentially hazardous elements in soil, environmental trajectories, risk assessments, and contamination controls. Further deepening his expertise, Professor Abbas delves into biochar studies, land restoration, and the intricate chemistry behind organic matter influencing soil and plant nutrition. Beyond his research, he lends his vast knowledge as a reviewer for an array of national and international academic journals.

Contents

Preface **XIII**

Section 1
Understanding Climate Change and Climate Smart Greenhouses **1**

Chapter 1 **3**
Introductory Chapter: Climate Change and Climate-Smart Greenhouses
by Ahmed A. Abdelhafez, Mohamed H.H. Abbas, Shawky M. Metwally, Hassan H. Abbas, Amera Sh. Metwally, Khaled M. Ibrahim, Aya Sh. Metwally, Rasha R.M. Mansour and Xu Zhang

Chapter 2 **19**
Hydroponic Production Systems in Greenhouses
by Božidar Benko, Sanja Fabek Uher, Sanja Radman and Nevena Opačić

Chapter 3 **45**
Agriculture's Contribution to the Emission of Greenhouse Gas Nitrous Oxide (N_2O) and Its Feasible Mitigation Strategies
by Raushan Kumar and Nirmali Bordoloi

Section 2
Innovations for Economic and Environmental Improvement **73**

Chapter 4 **75**
Innovative Greenhouse to Improve Economic and Environmental Conditions
by Zainab Abdel Mo'ez Mansour Embaby

Chapter 5 **87**
A Review of Controlled Environment Agriculture (CEA) Vegetable Production in Africa with Emphasis on Tomatoes, Onions and Cabbage
by Taiwo Bintu Ayinde, Charles Fredrick Nicholson and Benjamin Ahmed

Chapter 6 **99**
Simulation of a Novel Cooling System for a Closed Greenhouse
by Geordie Zapalac

Section 3
Assessing Environmental Impact and Mitigation Strategies **117**

Chapter 7 **119**
Life Cycle Assessment of Natural Gas Power Plant: Calculation of Impact Potentials
by Oludolapo Akanni Olanrewaju, Oluwafemi Ezekiel Ige, Busola Dorcas Akintayo and Ahad Ali

Preface

In the face of unprecedented challenges posed by climate change, the world of agriculture stands at a critical juncture. As global temperatures rise, weather patterns shift, and extreme events become more frequent, traditional farming methods are increasingly vulnerable. Recognizing the urgency of the situation, this book, *Climate Smart Greenhouses – Innovations and Impacts*, emerges as a timely exploration of cutting-edge technologies and sustainable practices that promise to revolutionize greenhouse agriculture.

The introductory chapter lays the groundwork, painting a vivid picture of the impact of climate change on agriculture. It sets the stage for a deep dive into the concept of climate-smart greenhouses, innovative solutions designed to adapt to and mitigate the challenges posed by a changing climate.

Hydroponic production systems take center stage in Chapter 2, revealing how this soilless cultivation method holds the key to resource-efficient and sustainable crop production within the confines of a greenhouse. Through practical examples and case studies, readers will gain insights into the transformative potential of hydroponics.

Chapter 3 addresses the environmental footprint of agriculture, focusing on nitrous oxide (N_2O) emissions. The discussion not only highlights the role of agriculture in greenhouse gas emissions but also provides feasible mitigation strategies. This chapter underscores the importance of adopting climate-smart practices to achieve a balance between food production and environmental stewardship.

Transitioning to the economic and environmental dimensions, Chapter 4 explores innovative greenhouse technologies. From energy-efficient designs to advanced monitoring systems, this chapter showcases how forward-thinking approaches can not only improve economic viability but also contribute to environmental sustainability.

Chapter 5 takes us to the African continent, providing a comprehensive review of Controlled Environment Agriculture (CEA) for vegetable production. With a spotlight on tomatoes, onions, and cabbage, this chapter explores how CEA practices can transform agriculture, addressing food security challenges and fostering economic development in Africa.

In Chapter 6, readers are introduced to a simulation of a novel cooling system tailored for closed greenhouse environments. The discussion goes beyond theory, presenting practical insights into the potential impact of such innovations on energy consumption and overall sustainability in greenhouse farming.

The final chapter conducts a life cycle assessment of a natural gas power plant, calculating impact potentials and emphasizing the delicate balance between energy production and greenhouse gas emissions. By addressing the environmental implications of energy choices in greenhouse operations, this chapter encourages readers to reflect on the interconnectedness of agriculture and broader environmental issues.

As we embark on this journey through *Climate Smart Greenhouses – Innovations and Impacts*, we hope that the diverse perspectives, innovative solutions, and practical insights contained within these pages will inspire a collective commitment to sustainable agriculture. This book is not just a compilation of chapters; it is a testament to the shared responsibility we all bear in nurturing our planet and securing a resilient future for generations to come.

May these pages catalyze change, spark conversations, foster collaboration, and pave the way for a greener, more sustainable future in greenhouse agriculture.

Warm regards!

Ahmed A. Abdelhafez
Faculty of Agriculture,
Department of Soils and Water,
New Valley University,
Egypt

National Committee of Soil Science,
Academy of Scientific Research and Technology,
Egypt

Mohamed H.H. Abbas
Faculty of Agriculture,
Department of Soils and Water,
Benha University,
Banha, Egypt

Section 1

Understanding Climate Change and Climate Smart Greenhouses

Chapter 1

Introductory Chapter: Climate Change and Climate-Smart Greenhouses

Ahmed A. Abdelhafez, Mohamed H.H. Abbas, Shawky M. Metwally, Hassan H. Abbas, Amera Sh. Metwally, Khaled M. Ibrahim, Aya Sh. Metwally, Rasha R.M. Mansour and Xu Zhang

1. Introduction

World is, nowadays, facing one of its most pressing ecological challenges — climate change. This phenomenon is characterized by significant and enduring shifts in weather patterns. It is mainly attributed to anthropogenic activities that increase the emissions of greenhouse gases [1] such as CO_2, CH_4, N_2O, O_3, chlorofluorocarbons (CFCs), CCl_4 [2], and H_2O [3]. Probably, CO_2 is the most important heat-trapping greenhouse gas (GHG) [4]. These gases absorb outgoing thermal (infrared) radiation, which is emitted by the surface of the Earth [5] and trap it within the atmosphere [6], thus increases the temperature of Earth's surface [7]. These gases also increase the temperature of troposphere while decreased stratosphere temperature [8]. In addition to GHGs, 5–15% of the organic carbon is found as aerosols that persist in the atmosphere for long time periods [9] and serve as cloud condensation nuclei (CCN) [10], which absorbs visible solar radiation (approximately 20% of total absorbed light) [11]. This is known by brown carbon, which are particulate matters containing chromophores that increase global warming threats [10]. The repercussions of climate change span a vast spectrum from altering weather patterns to impact human health and exacerbating disparities. This chapter delves into the intricate relationship between climate change and agriculture while spotlighting the innovative concept of climate-smart greenhouses as a promising solution.

2. Climate change: causes, impacts, and effects on human health

Climate change, a lasting alteration in weather patterns spanning decades to millions of years [6], results mainly from escalating greenhouse gas (GHG) emissions linked to activities such as fossil fuel combustion, deforestation, and industrial processes [12]. Carbon dioxide stands as the primary GHG from human activities [13], with methane and nitrous oxide trailing closely [14]. These gases trap heat in the Earth's atmosphere, elevating average global temperatures by 1–2% [15, 16].

These GHGs hasten the Earth's water cycle [8], leading to amplified evaporation from water surfaces [16], resulting in heightened drought frequency and intensity in various regions [17]. In contrast, other areas experience increased precipitation [16], leading to rising sea levels [12], expanding regional tides, and intensifying extreme events, such as hurricanes and floods [12]. Climate change exerts profound impacts on food and water supplies, human habitation, public health, and economic activities [18, 19]. The World Health Organization predicts that climate change may contribute to roughly 250,000 additional annual deaths by 2050, primarily due to malnutrition, malaria, diarrhea, and heat stress [20]. Vulnerable populations, including children, the elderly, low-income communities, and individuals with chronic illnesses, bear a disproportionate burden [19]. Heatwaves pose risks of heat exhaustion and heat stroke, exacerbating cardiovascular and respiratory diseases. Altered temperature and precipitation patterns also affect disease-carrying insects, increasing the transmission of illnesses, such as malaria and dengue fever [20]. Furthermore, climate change exerts adverse effects on mental health. The growing frequency and severity of extreme weather events and natural disasters contribute to post-traumatic stress disorder (PTSD), anxiety, depression, and other mental health issues [21]. Health disparities worsen with climate change, particularly impacting vulnerable populations with limited resources to adapt [22]. Addressing climate change necessitates a concerted global effort to reduce GHG emissions and adapt to ongoing changes [6].

3. The impact of climate change on agriculture

Climate change poses significant challenges to global agriculture and food security [12]. Rising temperatures, driven by the Industrial Revolution, have increased by 0.9°C since the nineteenth century and are projected to reach 2°C by 2100 [23, 24]. This warming can accelerate crop respiration and evapotranspiration [7], alter pest and disease distribution, and shorten the reproductive period in crops, such as wheat and rice [24, 25]. Wheat yields may decrease by 20–45%, and rice yields by 20–30% by 2100 [24]. Heat stress can cause post-heading carbon deficits in wheat [26]. Irregular precipitation patterns also impact agriculture [24], with some regions experiencing increased rainfall and others facing more frequent and severe droughts [7]. Both flooding and drought have adverse effects on crop yields, limiting growth, causing damage, and influencing the prevalence of pests and diseases [27, 28].

Climate change further challenges natural resource management in agriculture. Higher temperatures lead to increased evaporation rates, depleting water resources, and particularly affecting irrigation-dependent agriculture. Rising sea levels and increased salinity can degrade arable land, especially in coastal and delta regions [29]. To ensure agricultural sustainability in a changing climate, it is crucial to develop and implement adaptive strategies, including climate-resilient crop varieties, improved pest and disease management, and efficient water resource utilization [12].

4. Adaptation and mitigation strategies for climate change

Climate change presents significant challenges that require a comprehensive response. In 2015, an agreement was held in Paris to lessen the rise in global temperature by 2°C in 2100 [30]. Two primary strategies were adapted to attain this aim, which are adaptation and mitigation [31]. While mitigation focuses on reducing the causes of climate change, adaptation involves adjusting to its impacts [12]. Adaptation, therefore, is essential to manage unavoidable impacts, while mitigation is needed to limit the long-term changes in climate. The optimal mix of adaptation and mitigation measures will depend on local and regional factors, including climate change impacts, economic structures, and societal values.

5. Adaptation

Adaptation strategies mitigate global warming impacts [32] for future food security [33]. They involve altering processes, practices, and structures to reduce potential damage and capitalize on climate change opportunities [34]. Adaptation spans individual actions to institutional policies and includes financial adjustments [35]. In agriculture, strategies encompass crop enhancements, food waste reduction [33], water conservation policies [36], adjusted planting schedules to avoid extreme weather, and adopting resilient crop varieties [37]. Biodiversity enhancement can also aid climate mitigation [38]. Organic extracts, such as humic and fulvic acids or compost tea, promote plant growth, sequestering more CO_2 in plant tissues [39–41]. In the health sector, adaptation focuses on enhancing public health infrastructure to manage heatwaves and disease outbreaks, incorporating early warning systems, and improving air and water quality [42]. Urban areas should adapt by bolstering infrastructure resilience, implementing heat-wave action plans, creating cooling urban green spaces, and efficient water resource management [43].

Despite adaptation's importance in mitigating global warming's negative effects, maladaptation can occur [44]:

a. Infrastructural Maladaptation: For instance, sea walls in Fiji designed to combat rising sea levels hindered stormwater drainage. In Bangladesh, flood control measures reduced soil fertility and livelihood security.

b. Institutional Maladaptation: Farmers alter land use planning by growing cash crops, intercropping, and adopting moisture conservation techniques to address climate risks.

c. Behavioral Maladaptation: In northern Ghana, farmers migrate in search of employment, causing labor shortages.

6. Mitigation

Mitigation involves strategies to lessen or even stabilize emission of greenhouse gases in the atmosphere *via* adopting three strategies mentioned by Fawzy et al. [30] which are:

1. Switch to clean energy (renewable energy and nuclear power) or low carbon fuel to generate electricity and get heat rather than the burning of fossil fuels (conventional mitigation efforts). According to this approach, emissions of GHGs should be reduced by 45% in 2030 versus their levels in 2010 levels then reach finally to net-zero emissions by 2050.

2. Increasing the capability of sinks (oceans, soil, and forests) to capture and sequester GHGs gases (negative emissions technology).

3. Applying new techniques to managing solar and terrestrial radiation such as "stratospheric aerosol injection, marine sky brightening, cirrus cloud thinning, space-based mirrors and surface-based"; yet these methods are still theoretical.

Probably, energy production is the major source of greenhouse gas emissions, so applying new technologies to improve energy efficiency in buildings, transportations, and industrial processes might reduce effectively GHGs emissions [45]. In the agricultural sector, mitigation strategies include improving crop and livestock management practices to reduce methane and nitrous oxide emissions, protecting and restoring forests to sequester carbon, and managing soils to increase their carbon storage capacity [46, 47].

7. Climate-smart greenhouses: a sustainable approach to agriculture

The global agricultural sector faces climate change challenges, impacting crop yields and food security [12]. Climate-smart greenhouses, part of controlled environment agriculture (CEA), use advanced technologies to optimize plant growth and reduce environmental impact. They prioritize increasing productivity, adapting to climate change, and minimizing emissions [48, 49]. These greenhouses employ precision irrigation to conserve water and ensure ideal moisture levels, automated climate control to optimize growth and energy use, and energy-efficient lighting, such as LED, with minimal energy and water consumption [50–52]. Integrated pest management reduces reliance on chemical pesticides, benefiting crop health and sustainability [48].

Renewable energy sources, such as solar and wind power, further reduce environmental impact, and water conservation practices are employed [48]. Climate-smart greenhouses offer a promising solution to climate-related agricultural challenges, integrating technology and sustainable practices for increased productivity and reduced environmental impact [49].

8. Design and structure

Climate-smart greenhouses are designed to optimize the use of natural resources and energy. Sensors and devices (Internet of Things, IOT) can be used to monitor precisely, and then efficiently control all indoor parameters [52]. Greenhouses are often constructed with materials that maximize light transmission while provide insulation to reduce energy use for heating or cooling. Their structure and orientation allow maximizing natural light and regulating temperature [48]. Roofs and/or sides may be covered by an impermeable transparent plastic film to allow natural ventilation [53]. Android mobile applications are sometimes used to monitor and send warning messages about the state of plants.

9. Climate control systems

One of the key features of climate-smart greenhouses is to lessen energy consumption while increasing plant productivity [54]. This may take place *via* applying automated climate control systems that use different sensors to collect data continuously [50] and then use mathematical models for calculating solar irradiation, photosynthesis, and evapotranspiration [55]. These schemes can, therefore, regulate temperature, humidity, and light levels to create optimal growing conditions needed for plants at each growth stage to get high-yield production [56]. Ventilation and shading can also be adjusted to control the internal climate and reduce the need for artificial heating or cooling. In tropical and subtropical climates, covering materials are used for shading and cooling in greenhouses [56]. Some systems even include CO2 enrichment to enhance plant growth [48].

10. Water and nutrient management

Climate-smart greenhouses often incorporate precision irrigation and fertigation (fertilizer + irrigation) systems [57] using a combination of sensors and nutrient delivery schemes [58]. These systems deliver water and nutrients directly to the plant roots, reducing wastes, and ensuring that plants receive the optimal amount of moisture and nutrients. Some greenhouses also capture and reuse water through rainwater harvesting or condensation capture systems, contributing to water conservation efforts [48]. The emission of N gases from high-tech greenhouses that follow efficient recirculation systems is thought to be very low [59].

11. Energy efficiency and renewable energy

Energy efficiency is the key aspect of climate-smart greenhouses. Many climate-smart greenhouses use renewable energy sources, such as solar or wind power [48]. Using photovoltaic-thermal collectors of solar energy can produce both heat and electricity, with less shading [60]. Also, using energy-efficient lighting, such as LED lights, provides specific light spectrum needed for photosynthesis while using less electricity (40%) than traditional lighting systems and also less heat (9–49%) [61]. In cold regions, minimizing heating cost is another challenge. Thus, isolating greenhouses and/or using geothermal energy may help to lessen these costs [62].

12. Integrated pest management

Climate-smart greenhouses often use integrated pest management strategies to reduce the need for chemical pesticides. These strategies include the use of beneficial insects to control pests, use of physical barriers or traps, and the careful monitoring of pest populations to determine when control measures are needed [63, 64]. Microbial pesticides can also be used if natural enemies are not sufficient for pest control [65]. Moreover, solar ultraviolet-B lamps can provide a physical control for spider mites [66]. Climate-smart greenhouses represent a promising solution to the challenges posed by climate change in the agricultural sector. By integrating advanced technologies and sustainable practices, these greenhouses increase agricultural productivity, adapt to changing climate conditions, and reduce environmental impacts.

13. The role of greenhouse cultivation in climate change mitigation

Greenhouse cultivation, particularly when implemented with climate-smart practices, can play significant roles in mitigating climate change. This can be achieved *via* applying more efficient techniques in resource management, reducing wastes, and carbon sequestration. Generally, there are two methods to control greenhouse conditions (i) a passive method that depends on a natural phenomenon "hot air rises and cold air sinks," so it requires minimum energy while (ii) the active method needs fans for and heaters to control the environment inside greenhouses [67]. By means of thermal energy storage (TES) systems, heat can be successfully stabilized within greenhouses for plants [68]. These systems analyze the complex thermal processes within this indoor microclimate area and contribute toward efficient usage of this energy [69]. On the other hand, CO2 enrichment environment inside greenhouses can boost plant growth by approximately 35% [69, 70] *via* sequestrating CO2 from ambient air rather than being emitted to the atmosphere to increase the emissions of GHGs [71].

14. Efficient use of resources

Managing agricultural resources to meet rising food demands due to population growth is crucial [69], but natural resource limitations challenge food production [72]. Agricultural activities also contribute significantly to greenhouse gas (GHG) emissions [68], emphasizing the need for ecological considerations. Controlled environment agriculture (CEA), such as greenhouses, offers year-round food production possibilities [67]. Utilizing intelligent shading systems, smart glass, sensors, IoT, and AI [68], greenhouses precisely control conditions such as temperature, light, and humidity [73], increasing yield per unit area compared to traditional methods and conserving water through precision irrigation [48]. Some greenhouses capture and reuse water, further reducing water use [48]. Bioagents in greenhouses enhance horticultural yields and environmentally friendly pest and disease control, reducing GHG emissions related to agrochemicals [64, 74]. Greenhouses, by growing crops near consumption points, reduce transportation-related carbon emissions, especially in urban agriculture [75].

15. Carbon sequestration

Greenhouse gases can be reduced *via* a process known by phytosequestration [6]. In this method, plants absorb carbon dioxide gas from the atmosphere, change it to organic forms *via* Calvin cycle then sequester large amounts of C in their biomasses [76]. Herbaceous plants, which have relatively low planting-environment requirements, exhibit more capability to sequester C in their tissues than woody plants[77]. Surprisingly, sequestration of CO2 by microalgae is deemed as a net zero GHG emissions [78]. On the other hand, amounts of carbon sequestered *via* this process are relatively slow versus CO2 release due to anthropogenic activities [79]. Also, this process lasts for relatively short time periods because when plants decay and sequestered C returns back to air [6]. Weighing up pros and cons of phytosequestration, reforesting, and managing ecosystems are still effective ways to mitigate the global warming threat [79].

DOI: http://dx.doi.org/10.5772/intechopen.113212

16. Reduced waste and emissions

The terrestrial carbon pool is four times larger than the atmospheric carbon pool [4]. Recycling agricultural waste can significantly reduce greenhouse gas emissions, with over 5.6 billion mega grams of carbon potentially sequestered from the 18 billion wasted annually worldwide [80]. This can be achieved by converting organic residues into biochar, produced under limited oxygen conditions [81–86]. Biochar reduces easily oxidized carbon content, decreasing microbial metabolic activity by 47% [87], leading to longer soil retention when used as a soil amendment [88–90] or organic fertilizer [91, 92]. Its porous structure enhances soil CO2 adsorption *via* physisorption and chemisorption [93], sequestering carbon instead of releasing it into the atmosphere [87]. Scientists have devised a method to capture smoke emissions during biochar pyrolysis, using them for soil injection to improve seed germination and potentially achieve net-zero greenhouse gas emissions [94, 95]. Using biochar as a soil amendment can reduce CO2 emissions by about 1/8 [95], while converting residues to charcoal may cut GHG emissions by 80% within 8.5 years [96]. The potential for carbon sequestration in greenhouses remains an ongoing research topic, with outcomes depending on various factors, including greenhouse type, crop varieties, and management practice.

In conclusion, climate-smart greenhouse can contribute to climate change mitigation through more efficient use of resources, reduced waste and emissions, and potentially through carbon sequestration. However, it is important to note that not all greenhouses are the same, and the climate impact will depend on the specific design and management practices used.

Author details

Ahmed A. Abdelhafez[1,2*], Mohamed H.H. Abbas[3], Shawky M. Metwally[4], Hassan H. Abbas[3], Amera Sh. Metwally[5], Khaled M. Ibrahim[6], Aya Sh. Metwally[7], Rasha R.M. Mansour[8] and Xu Zhang[9]

1 Department of Soils and Water, Faculty of Agriculture, New Valley University, Egypt

2 National Committee of Soil Science, Academy of Scientific Research and Technology, Egypt

3 Department of Soils and Water, Faculty of Agriculture, Benha University, Egypt

4 Department of Soils and Water, Faculty of Technology and Development, Zagazig University, Egypt

5 Zagazig University Hospitals, Zagazig University, Zagazig, Egypt

6 Department of Agronomy, Faculty of Agriculture, New Valley University, Egypt

7 Department of Pharmacology, Faculty of Veterinary Medicine, Aswan University, Aswan, Egypt

8 Faculty of Specific Education, Benha University, Benha, Egypt

9 Eco-Environmental Protection Research Institute, Shanghai Academy of Agricultural Sciences, Shanghai, China

*Address all correspondence to: ahmed.aziz@agr.nvu.edu.eg

References

[1] Kabir M, Habiba U, Iqbal MZ, Shafiq M, Farooqi ZR, Shah A, et al. Impacts of anthropogenic activities & climate change resulting from increasing concentration of Carbon dioxide on environment in 21st Century; A Critical Review. IOP Conference Series: Earth and Environmental Science. 2023;**1194**:012010. DOI: 10.1088/1755-1315/1194/1/012010

[2] Sonwani S, Saxena P. Introduction to greenhouse gases: Sources, sinks and mitigation. In: Sonwani S, Saxena P, editors. Greenhouse Gases: Sources, Sinks and Mitigation. Singapore: Springer; 2022. pp. 1-7

[3] Thacker I, Sinatra GM. Visualizing the greenhouse effect: Restructuring mental models of climate Change through a guided online simulation. Education Sciences. 2019;**9**:14

[4] Yadav S, Sonkar V, Malyan SK. Soil carbon sequestration strategies: Application of biochar an option to combat global warming. In: Pandey VC, editor. Bio-Inspired Land Remediation. Environmental Contamination Remediation and Management. Cham: Springer; 2023

[5] Mosa A, Mansour MM, Soliman E, El-Ghamry A, El Alfy M, El Kenawy AM. Biochar as a soil amendment for restraining greenhouse gases emission and improving soil carbon sink: Current situation and ways forward. Sustainability. 2023;**15**:1206

[6] Jansson C, Wullschleger SD, Kalluri UC, Tuskan GA. Phytosequestration: Carbon biosequestration by plants and the prospects of genetic engineering. Bioscience. 2010;**60**:685-696. DOI: 10.1525/bio.2010.60.9.6

[7] Malhi GS, Kaur M, Kaushik P. Impact of climate Change on agriculture and its mitigation strategies: A review. Sustainability. 2021;**13**:1318

[8] Manabe S. Role of greenhouse gas in climate change. Tellus A: Dynamic Meteorology and Oceanography. 2019;**71**(1):1620078. DOI: 10.1080/16000870.2019.1620078

[9] Lopatka A. Brown carbon aerosols warm the air more than previously thought. Physics Today. 2023. DOI: 10.1063/PT.6.1.20230810a

[10] Shiraiwa M. Facing global climate and environmental change. ACS Environmental. 2023;**3**:121-122. DOI: 10.1021/acsenvironau.3c00014

[11] Wu G, Wan X, Ram K, Li P, Liu B, Yin Y, et al. Light absorption, fluorescence properties and sources of brown carbon aerosols in the southeast Tibetan plateau. Environmental Pollution. 2020;**257**:113616. DOI: 10.1016/j.envpol.2019.113616

[12] IPCC. Climate Change 2014: Impacts, adaptation, and vulnerability. Part A: Global and sectoral aspects. In: Metz B, Davidson OR, Bosch PR, Dave R, Meyer LA, editors. Contribution of Working Group II to the Fifth Assessment Report of the Intergovernmental Panel on Climate Change. Cambridge, United Kingdom: Cambridge University Press; 2014

[13] Anwar MN, Fayyaz A, Sohail NF, Khokhar MF, Baqar M, Yasar A, et al. CO_2 utilization: Turning greenhouse gas into fuels and valuable products. Journal of Environmental Management. 2020;**260**:110059. DOI: 10.1016/j.jenvman.2019.110059

[14] Kumar A, Singh P, Raizada P, Hussain CM. Impact of COVID-19 on greenhouse gases emissions: A critical review. Science Total Environment. 2022;**806**:150349. DOI: 10.1016/j.scitotenv.2021.150349

[15] Hoegh-Guldberg O, Jacob D, Taylor M, Guillén Bolaños T, Bindi M, Brown S, et al. The human imperative of stabilizing global climate change at 1.5°C. Science. 2019;**365**:eaaw6974. DOI: 10.1126/science.aaw6974

[16] Giorgi F, Raffaele F, Coppola E. The response of precipitation characteristics to global warming from climate projections. Earth System Dynamics. 2019;**10**:73-89. DOI: 10.5194/esd-10-73-2019

[17] Zandalinas S, Mittler R. Global warming, climate change, and environmental pollution: Recipe for a multifactorial stress combination disaster. Trends in Plant Science. 2021;**26**:588-599. DOI: 10.1016/j.tplants.2021.02.011

[18] Lin S, Zhang S, Yang Q, Cai Y, Li X, Ren Z. Rapid urbanization and global warming significantly impact tidal dynamics in the Pearl River estuary, China. Watershed Ecology and the Environment. 2023;**5**:100-107. DOI: 10.1016/j.wsee.2023.03.001

[19] Ahima RS. Global warming threatens human thermoregulation and survival. The Journal of Clinical Investigation. 2020;**130**:559-561. DOI: 10.1172/JCI135006

[20] WHO. Quantitative Risk Assessment of the Effects of Climate Change on Selected Causes of Death, 2030s and 2050s. Geneva, Switzerland: WHO; 2014

[21] Clayton S, Manning CM, Krygsman K, Speiser M. Mental Health and Our Changing Climate: Impacts, Implications, and Guidance. Washington, D.C.: American Psychological Association; 2020

[22] Watts N, Amann M, Arnell N, Ayeb-Karlsson S, Belesova K, Berry H, et al. The 2018 report of the lancet countdown on health and climate change: Shaping the health of nations for centuries to come. The Lancet. 2018;**392**:2479-2514. DOI: 10.1016/s0140-6736(18)32594-7

[23] Lobell DB, Gourdji SM. The influence of climate change on global crop productivity. Plant Physiology. 2012;**160**:1686-1697. DOI: 10.1104/pp.112.208298

[24] Arora NK. Impact of climate change on agriculture production and its sustainable solutions. Environmental Sustainability. 2019;**2**:95-96. DOI: 10.1007/s42398-019-00078-w

[25] Skendžić S, Zovko M, Živković IP, Lešić V. The impact of climate change on agricultural insect pests. Insects. 2021;**12**:440. DOI: 10.3390/insects12050440

[26] Shao L, Liu Z, Li H, Zhang Y, Dong M, Guo X, et al. The impact of global dimming on crop yields is determined by the source–sink imbalance of carbon during grain filling. Global Change Biology. 2021;**27**:689-708. DOI: 10.1111/gcb.15453

[27] Trenberth KE, Dai A, van der Schrier G, Jones PD, Barichivich J, Briffa KR, et al. Global warming and changes in drought. Nature Climate Change. 2014;**4**:17-22. DOI: 10.1038/nclimate2067

[28] Bebber DP, Ramotowski MA, Gurr SJ. Crop pests and pathogens move polewards in a warming world.

Nature Climate Change. 2013;**3**:985-988. DOI: 10.1038/nature.2013.13644

[29] Knox J, Hess T, Daccache A, Wheeler T. Climate change impacts on crop productivity in Africa and South Asia. Environmental Research Letters. 2012;**7**:034032. DOI: 10.1088/1748-9326/7/3/034032

[30] Fawzy S, Osman AI, Doran J, Rooney D. Strategies for mitigation of climate change: A review. Environmental Chemistry Letters. 2020;**18**:2069-2094. DOI: 10.1007/s10311-020-01059-w

[31] Abbass K, Qasim MZ, Song H, Murshed M, Mahmood H, Younis I. A review of the global climate change impacts, adaptation, and sustainable mitigation measures. Environmental Science and Pollution Research. 2022;**29**:42539-42559. DOI: 10.1007/s11356-022-19718-6

[32] Aryal JP, Sapkota TB, Khurana R, Khatri-Chhetri A, Lat ML. Climate change and agriculture in South Asia: Adaptation options in smallholder production systems. Environment, Development and Sustainability. 2020;**22**:5045-5075. DOI: 10.1007/s10668-019-00414-4

[33] Anderson R, Bayer PE, Edwards D. Climate change and the need for agricultural adaptation. Current Opinion in Plant Biology. 2020;**56**:197-202. DOI: 10.1016/j.pbi.2019.12.006

[34] Neil Adger W, Arnell NW, Tompkins EL. Successful adaptation to climate change across scales. Global Environmental Change. 2005;**15**:77-86. DOI: 10.1016/j.gloenvcha.2004.12.005

[35] Berrang-Ford L, Siders AR, Lesnikowski A, Fischer AP, Callaghan MW, Haddaway NR, et al. A systematic global stocktake of evidence on human adaptation to climate change. Nature Climate Change. 2021;**11**:989-1000. DOI: 10.1038/s41558-021-01170-y

[36] Gruda N, Bisbis M, Tanny J. Influence of climate change on protected cultivation: Impacts and sustainable adaptation strategies - a review. Journal of Cleaner Production. 2019;**225**:481-495. DOI: 10.1016/j.jclepro.2019.03.210

[37] Howden SM, Soussana J-F, Tubiello FN, Chhetri N, Dunlop M, Meinke H. Adapting agriculture to climate change. Proceedings of the National Academy of Sciences. 2007;**104**:19691-19696. DOI: 10.1073/pnas.0701890104

[38] Malhi Y, Franklin J, Seddon N, Solan M, Turner MG, Field CB, et al. Climate change and ecosystems: Threats, opportunities and solutions. Philosophical Transactions of the Royal Society B: Biological Sciences. 2020;**375**:20190104. DOI: 10.1098/rstb.2019.0104

[39] Farid IM, Abbas MH, et al. Organic materials and their chemically extracted humic and fulvic acids as potential soil amendments for Faba bean cultivation in soils with varying $CaCO_3$ contents. Horticulturae. 2021;**7**:205

[40] Farid I, Abbas M, El-Ghozoli A. Wheat productivity as influenced by integrated mineral, organic and biofertilization. Egyptian Journal of Soil Science. 2023;**63**:287-299. DOI: 10.21608/ejss.2023.192023.1590

[41] Farid IM, Abbas MHH, El-Ghozoli A. Implications of humic, fulvic and K—Humate extracted from each of compost and biogas manure as well as their teas on faba bean plants grown on Typic Torripsamments and emissions of soil CO2. Egyptian Journal of Soil

Science. 2018;**58**:275-293. DOI: 0.21608/ejss.2018.4232.1183

[42] Ebi KL, Semenza JC. Community-based adaptation to the health impacts of climate change. American Journal of Preventive Medicine. 2008;**35**:501-507. DOI: 10.1016/j.amepre.2008.08.018

[43] Rosenzweig C, Solecki WD, Hammer SA, Mehrotra S, editors. Climate Change and Cities: First Assessment Report of the Urban Climate Change Research Network. Kingdom of England: Cambridge University Press; 2011

[44] Schipper ELF. Maladaptation: When adaptation to climate change goes very wrong. On Earth. 2020;**3**:409-414. DOI: 10.1016/j.oneear.2020.09.014

[45] Clarke L, Jiang K, Akimoto K, Babiker M, Blanford G, Fisher-Vanden K, et al. Assessing transformation pathways. In: Climate Change 2014: Mitigation of Climate Change: Working Group III Contribution to the IPCC Fifth Assessment Report, Intergovernmental Panel on Climate, C. Cambridge: Cambridge University Press; 2015. pp. 413-510

[46] Smith P, Martino D, Cai Z, Gwary D, Janzen H, Kumar P, et al. Agriculture in climate Change 2007: Mitigation. In: Metz B, Davidson OR, Bosch PR, Dave R, Meyer LA, editors. Contribution of Working Group III to the Fourth Assessment Report of the Intergovernmental Panel on Climate Change. Cambridge, United Kingdom: Cambridge University Press; 2007

[47] Memarbashi P, Mojarradi G, Keshavarz M. Climate-smart agriculture in Iran: Strategies, constraints and drivers. Sustainability. 2022;**14**:15573. DOI: 10.3390/su142315573

[48] Kozai T, Niu G, Takagaki M. Plant Factory: An Indoor Vertical Farming System for Efficient Quality Food Production. Cambridge: Academic Press; 2015

[49] FAO. Climate-Smart Agriculture Sourcebook. Rome, Italy: FAO; 2013

[50] Bersani C, Ruggiero C, Sacile R, Soussi A, E., Z. Internet of things approaches for monitoring and control of smart greenhouses in industry 4.0. Energies. 2022;**15**:3834. DOI: 10.3390/en15103834

[51] Cecchetti G, Ruscelli AL. Monitoring and automation for sustainable smart greenhouses. In: 2022 IEEE International Conference on Smart Computing (SMARTCOMP). Helsinki, Finland; 2022. pp. 381-386. DOI: 10.1109/SMARTCOMP55677.2022.00084

[52] Bersani C, Ouammi A, Sacile R, E., Z. Model predictive control of smart greenhouses as the path towards near zero energy consumption. Energies. 2020;**13**:3647. DOI: 10.3390/en13143647

[53] Gruda N, Bisbis M, Tanny J. Impacts of protected vegetable cultivation on climate change and adaptation strategies for cleaner production – A review. Journal of Cleaner Production. 2019;**225**:324-339. DOI: 10.1016/j.jclepro.2019.03.295

[54] Maher A, Kamel E, Enrico F, Atif I, Abdelkader M. An intelligent system for the climate control and energy savings in agricultural greenhouses. Energy Efficiency. 2016;**9**:1241-1255. DOI: 10.1007/s12053-015-9421-8

[55] Aaslyng JM, Lund JB, Ehler N, Rosenqvist E. IntelliGrow: A greenhouse component-based climate control system. Environmental Modelling &

Software. 2003;**18**:657-666. DOI: 10.1016/S1364-8152(03)00052-5

[56] Choab N, Allouhi A, El Maakoul A, Kousksou T, Saadeddine S, Jamil A. Review on greenhouse microclimate and application: Design parameters, thermal modeling and simulation, climate controlling technologies. Solar Energy. 2019;**191**:109-137. DOI: 10.1016/j.solener.2019.08.042

[57] Gallardo M, Elia A, Thompson RB. Decision support systems and models for aiding irrigation and nutrient management of vegetable crops. Agricultural Water Management. 2020;**240**:106209. DOI: 10.1016/j.agwat.2020.106209

[58] Massa D, Magán JJ, Montesano FF, Tzortzakis N. Minimizing water and nutrient losses from soilless cropping in southern Europe. Agricultural Water Management. 2020;**241**:106395. DOI: 10.1016/j.agwat.2020.106395

[59] Zhou D, Meinke H, Wilson M, Marcelis LFM, Heuvelink E. Towards delivering on the sustainable development goals in greenhouse production systems. Resources, Conservation and Recycling. 2021;**169**:105379. DOI: 10.1016/j.resconrec.2020.105379

[60] Gorjian S, Calise F, Kant K, Ahamed MS, Copertaro B, Najafi G, et al. A review on opportunities for implementation of solar energy technologies in agricultural greenhouses. Journal of Cleaner Production. 2021;**285**:124807. DOI: 10.1016/j.jclepro.2020.124807

[61] Katzin D, Marcelis LFM, van Mourik S. Energy savings in greenhouses by transition from high-pressure sodium to LED lighting. Applied Energy. 2021;**281**:116019. DOI: 10.1016/j.apenergy.2020.116019

[62] Ahamed MS, Guo H, Tanino K. Energy saving techniques for reducing the heating cost of conventional greenhouses. Biosystems Engineering. 2019;**178**:9-33. DOI: 10.1016/j.biosystemseng.2018.10.017

[63] Daughtrey M, Buitenhuis R. Integrated Pest and disease Management in Greenhouse Ornamentals. In: Gullino ML, Albajes R, Nicot PC, editors. Integrated Pest and Disease Management in Greenhouse Crops. Cham: Springer International Publishing; 2020. pp. 625-679

[64] Dhawan AK, Peshin R. Integrated Pest management: Concept, opportunities and challenges. In: Peshin R, Dhawan AK, editors. Integrated Pest Management: Innovation-Development Process. Vol. 1. Dordrecht: Springer Netherlands; 2009. pp. 51-81

[65] van Lenteren JC, Alomar O, Ravensberg WJ, Urbaneja A. Biological control agents for control of pests in greenhouses. In: Gullino ML, Albajes R, Nicot PC, editors. Integrated Pest and Disease Management in Greenhouse Crops. Cham: Springer International Publishing; 2020. pp. 409-439

[66] Osakabe M. Biological impact of ultraviolet-B radiation on spider mites and its application in integrated pest management. Applied Entomology and Zoology. 2021;**56**:139-155. DOI: 10.1007/s13355-020-00719-1

[67] Fox JA, Adriaanse P, Stacey NT. Greenhouse energy management: The thermal interaction of greenhouses with the ground. Journal of Cleaner Production. 2019;**235**:288-296. DOI: 10.1016/j.jclepro.2019.06.344

[68] Maraveas C, Karavas C-S, Loukatos D, et al. Agricultural greenhouses: Resource management technologies and perspectives for zero greenhouse gas emissions. Agriculture. 2023;**13**:1464. DOI: 10.3390/agriculture13071464

[69] Iddio E, Wang L, Thomas Y, McMorrow G, Denzer A. Energy efficient operation and modeling for greenhouses: A literature review. Renewable and Sustainable Energy Reviews. 2020;**117**:109480. DOI: 10.1016/j.rser.2019.109480

[70] Gamage D, Thompson M, Sutherland M, Hirotsu N, Makino A, Seneweera S. New insights into the cellular mechanisms of plant growth at elevated atmospheric carbon dioxide concentrations. Plant, Cell & Environment. 2018;**41**:1233-1246. DOI: 10.1111/pce.13206

[71] Bao J, Lu W-H, Zhao J, Bi XT. Greenhouses for CO_2 sequestration from atmosphere. Carbon Resources Conversion. 2018;**1**:183-190. DOI: 10.1016/j.crcon.2018.08.002

[72] Wang X. Managing land carrying capacity: Key to achieving sustainable production systems for food security. Land. 2022;**11**:484. DOI: 10.3390/land11040484

[73] Tawalbeh M, Aljaghoub H, Alami AH, Olabi AG. Selection criteria of cooling technologies for sustainable greenhouses: A comprehensive review. Thermal Science and Engineering Progress. 2023;**38**:101666. DOI: 10.1016/j.tsep.2023.101666

[74] Li H, Wyckhuys KAG, Wu K. Hoverflies provide pollination and biological pest control in greenhouse-grown horticultural crops. Frontiers in Plant Science. 2023;**14**:1118388. DOI: 10.3389/fpls.2023.1118388

[75] Despommier D, editor. The Vertical Farm: Feeding the World in the 21st Century. New York City, United States: Macmillan-London, United Kingdom and Equitable Building; 2010

[76] Lehmann J. A handful of carbon. Nature. 2007;**447**:143-144. DOI: 10.1038/447143a

[77] Deng L, Yuan H, Xie J, Ge L, Chen Y. Herbaceous plants are better than woody plants for carbon sequestration. Resources, Conservation and Recycling. 2022;**184**:106431. DOI: 10.1016/j.resconrec.2022.106431

[78] Ma Z, Cheah WY, Ng I-S, Chang J-S, Zhao M, Show PL. Microalgae-based biotechnological sequestration of carbon dioxide for net zero emissions. Trends in Biotechnology. 2022;**40**:1439-1453. DOI: 10.1016/j.tibtech.2022.09.002

[79] Baldocchi D, Penuelas J. The physics and ecology of mining carbon dioxide from the atmosphere by ecosystems. Global Change Biology. 2019;**25**:1191-1197. DOI: 10.1111/gcb.14559

[80] Marxsen CS. Potential world garbage and waste carbon sequestration. Environmental Science & Policy. 2001;**4**:293-300. DOI: 10.1016/S1462-9011(01)00035-1

[81] Lalarukh I, Amjad SF, Mansoora N, Al-Dhumri SA, Alshahri AH, Almutari MM, et al. Integral effects of brassinosteroids and timber waste biochar enhances the drought tolerance capacity of wheat plant. Scientific Reports. 2022;**12**:12842. DOI: 10.1038/s41598-022-16866-0

[82] Asaad AA, El-Hawary AM, Abbas MHH, Mohamed I, Abdelhafez AA, Bassouny MA. Reclamation of wastewater in wetlands using reed plants and biochar. Scientific Reports.

2022;**12**:19516. DOI: 10.1038/s41598-022-24078-9

[83] Farid IM, Siam HS, Abbas MHH, Mohamed I, Mahmoud SA, Tolba M, et al. Co-composted biochar derived from rice straw and sugarcane bagasse improved soil properties, carbon balance, and zucchini growth in a sandy soil: A trial for enhancing the health of low fertile arid soils. Chemosphere. 2022;**292**:133389. DOI: 10.1016/j.chemosphere.2021.133389

[84] Khalil FW, Abdel-Salam M, Abbas MHH, Abuzaid AS. Implications of acidified and non-acidified biochars on N and K availability and their uptake by maize plants. Egyptian Journal of Soil Science. 2023;**63**:101-112. DOI: 10.21608/ejss.2023.184654.1560

[85] Bassouny M, Abbas M. Role of biochar in managing the irrigation water requirements of maize plants: The pyramid model signifying the soil hydro-physical and environmental markers. Egyptian Journal of Soil Science. 2019;**59**:99-115. DOI: 10.21608/ejss.2019.9990.1252

[86] Tolba M, Farid I, Siam H, Abbas M, Mohamed I, Mahmoud S, et al. Integrated management of K -additives to improve the productivity of zucchini plants grown on a poor fertile sandy soil. Egyptian Journal of Soil Science. 2021;**61**:355-365. DOI: 10.21608/ejss.2021.99643.1472

[87] Liu H, Wang X, Song X, Leng P, Li J, Rodrigues JLM, et al. Generalists and specialists decomposing labile and aromatic biochar compounds and sequestering carbon in soil. Geoderma. 2022;**428**:116176. DOI: 10.1016/j.geoderma.2022.116176

[88] Abdelhafez AA, Abbas MHH, Li J. Biochar: the black diamond for soil sustainability, contamination control and agricultural production. In: Engineering Applications of Biochar. London, UK, London, UK: IntechOpen; 2017. pp. 7-27

[89] Abdelhafez AA, Zhang X, Zhou L, Zou G, Cui N, Abbas MHH, et al. Introductory Chapter: Is Biochar Safe? Vol. 2020. London, UK, London, UK: IntechOpen; 2020. pp. 1-6

[90] Zhang X, Zou G, Chu H, Shen Z, Zhang Y, Abbas MHH, et al. Biochar applications for treating potentially toxic elements (PTEs) contaminated soils and water: A review. Frontiers in Bioengineering and Biotechnology. 2023;**11**. DOI: 10.3389/fbioe.2023.1258483

[91] Lehmann J, Cowie A, Masiello CA, Kammann C, Woolf D, Amonette JE, et al. Biochar in climate change mitigation. Nature Geoscience. 2021;**14**:883-892. DOI: 10.1038/s41561-021-00852-8

[92] Elshony M, Farid I, Alkamar F, Abbas MHH, Abbas H. Ameliorating a sandy soil using biochar and compost amendments and their implications as slow release fertilizers on plant growth. Egyptian Journal of Soil Science. 2019;**59**:305-322. DOI: 10.21608/ejss.2019.12914.1276

[93] Ben Salem I, El Gamal M, Sharma M, Hameedi S, Howari FM. Utilization of the UAE date palm leaf biochar in carbon dioxide capture and sequestration processes. Journal of Environmental Management. 2021;**299**:113644. DOI: 10.1016/j.jenvman.2021.113644

[94] Abdelhafez AA, Zhang X, Zhou L, Cai M, Cui N, Chen G, et al. Eco-friendly production of biochar via conventional pyrolysis: Application of biochar and

liquefied smoke for plant productivity and seed germination. Environmental Technology & Innovation. 2021;**22**:101540. DOI: 10.1016/j.eti.2021.101540

[95] Kurniawan TA, Othman MHD, Liang X, Goh HH, Gikas P, Chong K-K, et al. Challenges and opportunities for biochar to promote circular economy and carbon neutrality. Journal of Environmental Management. 2023;**332**:117429. DOI: 10.1016/j.jenvman.2023.117429

[96] Thakkar J, Kumar A, Ghatora S, Canter C. Energy balance and greenhouse gas emissions from the production and sequestration of charcoal from agricultural residues. Renewable Energy. 2016;**94**:558-567. DOI: 10.1016/j.renene.2016.03.087

Chapter 2

Hydroponic Production Systems in Greenhouses

Božidar Benko, Sanja Fabek Uher, Sanja Radman and Nevena Opačić

Abstract

Hydroponic production means the growing of vegetables, herbs and ornamental plants and fruits in a nutrient solution (a solution of water and macro- and micronutrients) with or without the use of a substrate that gives the mechanical support to plant. The most important advantages of hydroponics are as follows: continuous cultivation of one crop, better control and supply of plants with water and plant nutrients, reduced occurrence of plant pests and minimized environmental impact and increased water use efficiency. The main hydroponic cultivation technique of fruit vegetables is cultivation on substrates, often called soilless system. Growing substrate (organic, inorganic or synthetic) provides an aseptic environment, good oxygenation and an adequate nutrient solution flow, so the most important substrate properties are biological and chemical inert, porosity and capillarity. Its choice depends on climatic conditions, the type of equipment in the greenhouse and the plant requirements. Hydroponics is also suitable for growing crops with a shorter growing period such as leafy vegetables and herbs. Plants are grown by different growing techniques in a nutrient solution without a substrate (nutrient film technique, floating hydroponics, ebb and flow and aeroponics). These are closed hydroponic systems, which means that drainage nutrient solution is collected, sterilized and reused.

Keywords: soilless culture, nutrient solution, inert substrates, water culture, open and closed systems, aeroponics, floating hydroponics fruit vegetables, leafy vegetables

1. Introduction

Human population increasing and market demands require major adjustments in the way food is produced, and turning from previous traditional forms of cultivation to new and sustainable ones. One way to increase the food production sustainability is to grow plants in different hydroponic production systems. Hydroponics represents a climate-smart production method, due to environmental concerns, resource sustainability and efficient use as well as climate changes [1, 2]. This mean, when compared to the open field production, which is often exposed to biotic and abiotic stress factors that hinder production, hydroponics use less resources such as land space, pesticide, and water. As in the most cases hydroponics are placed in the greenhouses, the control of production factors such as temperature, relative humidity, light and carbon dioxide, as well as extension of the growing season is possible. These production systems also make the supply and distribution of nutrients to crops easier and more uniform to enhance crop growth and yield [3].

Hydroponically grown plants in greenhouses are optimally supplied with water and nutrients and have optimal growth and development conditions due to climate control. Production mostly takes place in heated greenhouses, which allows the production and supply of the market throughout or most of the year, depending on the culture grown. Vegetables, herbs and ornamental plants and fruits are grown in a nutrient solution (solution of water and macro- and micronutrients) with or without the use of substrate that gives the mechanical support to plant. Plant nutrients are in optimal relation, and concentration determined by the electrical conductivity (EC-value) and the pH value.

Mentioned above results are with the advantages of hydroponics [2, 4]:

- plant cultivation in locations and areas where there is no soil or the soil is unsuitable for growing,
- continuous cultivation of one crop in the same production area (no crop rotation required),
- better control and supply of plants with water regarding time and amount,
- better control and supply of plants with plant nutrients (during the growing season, the concentration, composition, time and amount of nutrient solution are changed as needed, depending on the plant development phase and on the microclimatic conditions of the greenhouse),
- reduced occurrence of plant pests (diseases, pests, nematodes and weeds) that need greenhouse soil for their development and overwintering,
- minimized environmental impact and increased water use efficiency with closed hydroponic systems.

These advantages result in higher production of biomass in the time and area unit in hydroponic cultivation compared to the soil cultivation, and thus earlier harvesting (faster entry into technological maturity), more harvests in crops that multiple harvested and higher total yields. Besides that, hydroponics represents an appropriate and sustainable growing technology for urban and peri-urban areas, where higher yield could be achieved by using vertical space (vertical farming systems) to meet food demands in densely populated areas [5].

Disadvantages of hydroponic cultivation techniques are high initial investments, that is, higher costs of installing hydroponic systems in relation to conventional soil cultivation. Successful hydroponic production requires a high degree of knowledge and expertise in the field of agronomy and technical skills and knowledge to manage the equipment applied. If diseases and pests occur, the infection spreads rapidly due to optimal conditions for their development in a greenhouse. Due to significantly higher costs, the successful application of hydroponic technology is limited to species of high economic value, in some regions often to a certain part of the year. An additional problem of hydroponic cultivation techniques on substrates is also a disposal and recycling of inorganic and synthetic substrates after use [2, 4].

Hydroponic production systems include both cultivation on different inert substrates or growing media (soilless culture) and water culture with nutrient solution as root environment (without substrate). Regarding drainage solution usage hydroponic systems could be divided to open or closed. In open systems, the drainage solution is discharged, while in closed systems the drainage solution is collected, sterilized and reused [6]. Velasquez-Gonzales et al. [2] stated that choose of hydroponic growing technique depends on the plant species, local climate and budget, among other factors.

Despite some disadvantages mentioned above, hydroponic production is a rapidly growing sector that has seen tremendous growth in recent years. According to various statistics, the global hydroponic system market is projected to reach 16.03 billion USD by 2028 and Europe represents the largest market for this industry, accounting for 41% of its share. The compound annual growth rate of the hydroponics between 2022 and 2028 is estimated at 11.3%. Hydroponic greenhouse vegetable production is growing at a rate of 5–10% annually worldwide, and tomato is the most popular crop in the commercial hydroponics, accounting for over 30% of hydroponic production. When using hydroponic production systems, producers could achieve up to 2–4 times higher yield with approximately 90% less water consumed than traditional soil-based agriculture. At the same time, environmental pollution is decreased by nearly 70% [7].

To reach positive financial results, hydroponic production systems should be placed in a well-equipped, high-tech greenhouses where soilless culture equipment represents only a small fraction of the total investment of about 200 $€\cdot m^{-2}$. However, low- or mid-tech greenhouses may sometimes be modernized and used for hydroponics, depending on the economic and technical conditions, such as region, farm characteristics, type of greenhouse, soil problems, water resources, market requirements, establishment costs and, last but not least, restrictions on environment pollution. Low-cost alternatives are suitable for growers with limited capital or in regions with a fluctuating demand. In low-tech hydroponics, the heart of the system is the growing medium or water, while a simple system controls and distributes the nutrient solution or a drip irrigation system can be used [8].

This chapter will discuss about growing technology and substrates used in soilless culture and about water culture growing techniques.

2. Soilless culture

2.1 Growing substrate

There is no material or mixture that would be universal for growing all crops in all growing conditions. Growing substrate properties should match the requirements of the crop and the growing technology. Substrate should provide an aseptic environment, good oxygenation and an adequate nutrient solution flow, so the most important substrate properties are biological and chemical inert, porosity and capillarity. Biologically inert substrate means the absence of pathogens of plant diseases, pests and weed seeds. The chemical inert substrate does not contain any nutrients and does not affect, and they do not change the composition of the applied nutrient solution. In last years, life cycle and substrate sustainability (economic, social and environmental viability) becomes more and more important properties. For sustainable production of vegetables in growing media, priority should be given to locally available and not very

Figure 1.
Inorganic growing substrates (A: rockwool; B: perlite; C: expanded clay).

expensive or locally manufactured and standardized products. The choice of substrate as a growing medium depends on climatic conditions, the type of equipment in the greenhouse and the requirements of the plants that need to be met [2–6, 8].

Substrate	Volume weight, kg/m^3	Total porosity, vol. %	Water porosity, vol. %	Air porosity, vol. %	pH value
Sand	1400-1600	40–50	20–40	10–20	6.4–7.2
Pumice	570–630	80–90	2–5	75–85	7.0–8.0
Expanded clay	300–700	40–50	5–10	30–40	4.5–9.0
Perlite	90–130	50–75	15–35	30–60	6.5–7.5
Rockwool	55–90	95–97	75–80	10–15	7.0–7.5
Peat	60–400	55–97	52–88	6–42	3.0–7.3
Coconut fiber	65–110	94–96	80–85	10–12	5.0–6.8

Table 1.
Physical and chemical characteristics of some growing substrates [9, 10].

Hydroponic growing substrates are divided into organic (peat, coconut fiber, sawdust, corn, straw), inorganic (rockwool, perlite, sand, expanded clay, pumice, vermiculite, zeolite) and synthetic derived from petroleum (polystyrene, polyurethane and urea-formaldehyde foam). Also, they are divided into fibrous and granular. Fibrous are characterized by a high fiber content of different dimensions giving the substrate high water capacity and low for air. Retained water is easily accessible to the plant, and the volume is significantly reduced and varies from 2 to 7 liters per plant. Granulated substrates (sand, perlite) as opposed to fibrous ones have increased air capacity and reduced water by 10 to 40%. Retained water is more difficult to access to the plant, and the volume of substrate for one plant must be much higher than the fibrous substrates and amounts to between 10 and 40 liters [9].

2.1.1 Rockwool

Rockwool is a natural material obtained by heat treatment of volcanic rocks (bazalt and diabaz), which, with the addition of coke and limestone, are talent and refined to the final product, which, under the influence of high temperatures, acquires a fibrous structure. These fibers are then pressed into blocks or cubes (**Figure 1**) of light volume weight (80–90 kg·m^{-3}) [6]. It absorbs water very well and has good drainage properties. Total porosity is from 95 to 97%. Of these, 75 to 80% are water micropores and 10 to 15% are air macropores (**Table 1**). One of the most significant features of rockwool is its sterility, that is, the complete absence of pathogenic microorganisms and everything else that could contaminate soilless cultivation. It has mild alkaline reaction (pH value from 7 to 8.5). Because rockwool is an inert pH value can be easily reduced to optimal in hydroponic cultivation (from 6 to 6.5) using a slightly acidic nutrient solution. After use, it can be thermally sterilized and reused for one or 2 years, which reduces environmental pollution. However, after each use, the fiber structure worsens and reduces the proportion of air pores. In areas with colder climates, less density rockwool with vertical threads is most used, while for warmer appetizers, higher-density stone wool with horizontal threads is recommended to allow for better water retention [9, 10].

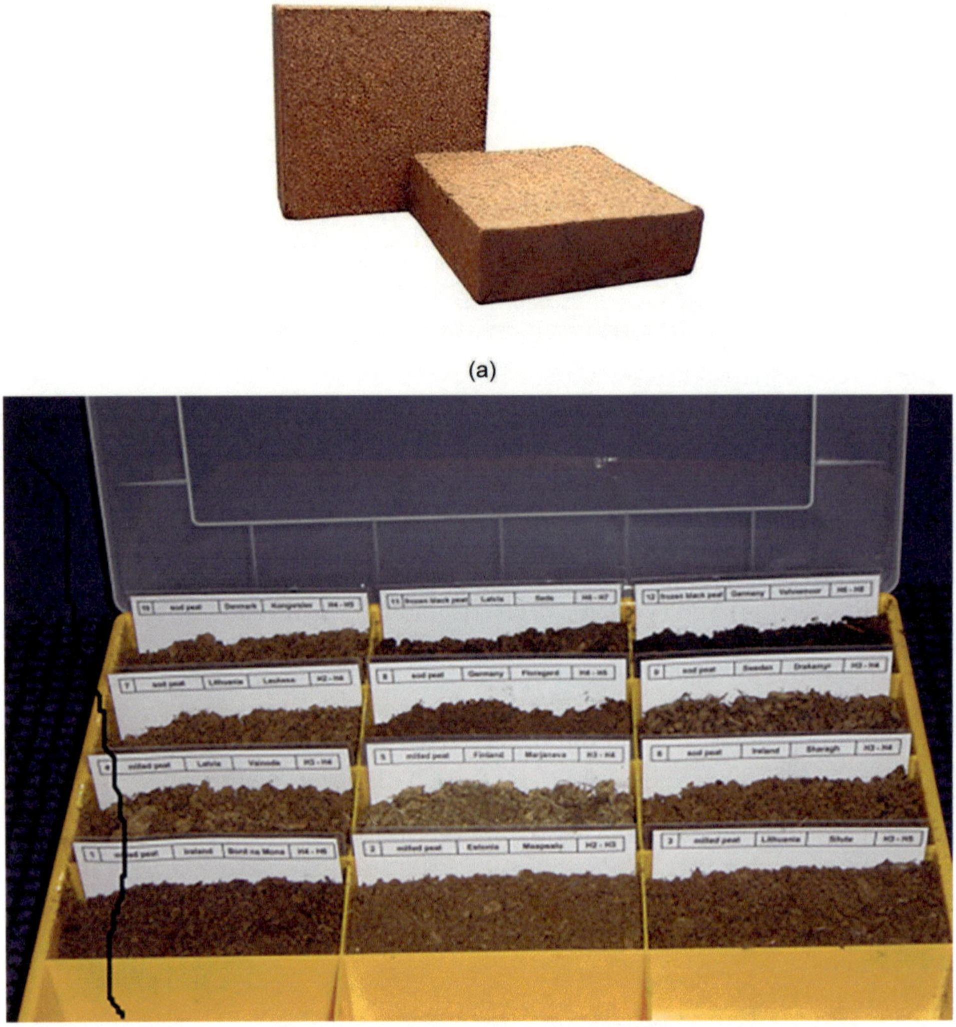

(a)

(b)

Figure 2.
Organic substrates: coconut fibers (a), and peat (b).

2.1.2 Perlite

Perlite (**Figure 1**) is an aluminum silicate of volcanic origin containing 75% SiO_2 and 13% Al_2O_3. It is a sterile material, neutral in pH (6.5–7.5) and no decay, with light volume weight (90–130 kg·m^{-3}) [6]. Its porosity (50–75%) ensures good breathability important for the growth of the root system (**Table 1**). Several different perlite granulations are produced (<3 mm, <5 mm, etc.). It can be purchased on the market packed in bags of volume 10 to 15 L on which the plants are planted or in large bags of volume about 100 L when filled into breeding vessels. It can be used alone or in a mixture with other substrates. If it is represented in a higher ratio in the mixture, attention should be paid to the pH value, which should not be lower than 5. It is often mixed with organic materials (peat) that improve its elasticity, permeability and other physical characteristics [9, 10].

The main disadvantage of rockwool and perlite is high energy consumption during production and their high price [6].

2.1.3 Expanded clay

Expanded clay (**Figure 1**) is obtained by roasting natural clay at 1200°C over 3 hours, giving a porous medium in the form of balls with a diameter of 4 to 20 mm, depending on the purpose. It is an inert substrate without a nourishing, neutral pH reaction. Capillarity on the surface of the ball provides a nutrient solution near the roots of the system. The balls dry easily and do not contain excess water that provides enough oxygen near the roots The lack of expanded clay is a fairly large volume mass, making it difficult to manipulate and very low water porosity (**Table 1**), which requires frequent and short fertigation [9, 10].

2.1.4 Coconut fiber

Coconut fiber (**Figure 2**) is increasingly used in hydroponic cultivation, and can be found on the market under different names, most often in the form of pressed blocks (plate). This substrate combines high water capacity of vermiculite and air capacity of perlite. However, it is completely organic in origin obtained by peeling coconuts. By its physical properties (**Table 1**), it is most similar to rockwool. Coconut fiber has physical stability [6], light weight (65–110 kg·m^{-3}), good air content, high total pore space between 94 and 96%, and water holding capacity, subacid-neutral pH (5–6.8). It is rich in hormones and sterilized by pressurized water vapor, which ensures ideal conditions for rooting and protects against the causes of plant diseases. Also, unlike peat, coconut fiber is a completely renewable resource. The lack of coconut fiber as a substrate can be the content of NaCl [6], which affects the ion concentration in the root zone and has a detrimental effect on its development. Pressed blocks require soaking in aqueous solution before use. During soaking, the substrate rehydration and swelling occur up to six times the initial size. It is very often mixed with perlite or vermiculite in equal proportions [9, 10].

2.1.5 Peat

Peat is the most important material of organic origin and is obtained from the remains of *Sphagnum* moss (**Figure 2**). It is characterized by good drainage and structure, that is, physical stability, good air and water holding capacity with total pore space ranging 85–97%, low microbial activity, light volume weight (60–200 kg m^{-3}), low and easily to adjusted pH, and low nutrient content [6]. According to the degree of decay, the amount of hinges is divided into white, brown and black peat. White peat has great absorption power and high acidity, and pH values between 3.5 and 4.2 (**Table 1**). It contains very few nutrients so it improves the water regime and air capacity. Black peat contains larger amounts of minerals that are suitable for plant growth. The reaction varies from 6.5 to 7.2 so that it is suitable for growing plants to suit a neutral or weakly alkaline reaction [9, 10].

Disadvantages of peat are that it is finite resource, environmental concerns and contribution to CO_2 release due to peatlands use, increasing cost due to energy crisis. It may be strongly acidic; shrinking may lead to substrate hydro-repellence [6].

Figure 3.
Tomato plants in rockwool plugs ready for pricking.

2.2 Soilless growing technology

The main form of fruit vegetables production is cultivation on substrates. The substrate is a medium whose role is to strengthen the root system, maintain water in the form of accessible plants, runoff of excess nutrients, and ensure air exchange. Soilless culture of fruit vegetables on substrates is technologically similar to soil cultivation in the greenhouse.

2.2.1 Greenhouse preparation

Before planting, it is necessary to prepare a greenhouse. In the greenhouse, equipment for nutrient solution preparation and a drip irrigation system should be installed. Substrate plates are placed in rows or double-row strips. The substrate is placed on hanging gutters, which serve to runoff an excess nutrient solution. The distance between the rows is 120 to 150 cm. If planted in double-row strips, the distance between the rows in the strip is 70 to 80 cm, and between the strips 100 to 120 cm. If cubes with two prickled plants are planted, two rows of plants are obtained from one row of substrate. After the substrate is placed, the planting sites are cut at the polyethylene foil into which the substrate is packaged. The distance between the plants in the row is 33 to 50 cm. Capillary carriers are inserted vertically into the cut openings so the substrate could be soaked with a nutrient solution before planting.

2.2.2 Sowing and planting

The seed sowing is most often done in rockwool plugs and planting on a selected inert substrate. Another possibility is sowing in rockwool blocks, 2.5-cm brides and

Figure 4.
Tomato seedlings in rockwool cubes.

Figure 5.
Cucumber planted in peat bags.

4 cm high. Fifty to sixty blocks are connected by an upper edge so that they form a larger sowing unit. The plugs are placed in polystyrene containers with 240 pots. Sowing is most often done in late November or early December. After sowing in

Figure 6.
Tomato plants at the beginning of harvest.

rockwool plugs, seeds are covered with vermiculite, which keeps constant temperature and retains moisture needed for emergence.

Emerged plants are pricked at the phase of developed cotyledon leaves and the first true leaf (**Figure 3**) into the rockwool cubes. The cube size depends on the culture and the number of plants being prickled into one cube. If one plant is prickled per cube, 7.5-cm edge cubes and 6.5 cm high or 10-cm edge cubes and 7.5 cm high are used, and if two plants are pricked per cube, cubes with 10- or 12-cm edge are used. Since seedlings are produced during a short day, supplemental lighting should be used in order to shorten the growing period.

Seedlings are grown in rockwool cubes until planting. During cultivation, they are fertigated with a nutrient solution of reduced concentration every day or every other day. If necessary, after watering with the solution, the leaves are rinsed with tap water to wash out the remains of nutrient salts. When the plants begin to touch each other, the cubes need to be spaced apart to prevent the seedling elongation. The cube is separated once and twice during the cultivation of seedlings (**Figure 4**). The seedlings are ready for planting when the root grows through the volume of the cube, that is, in late January or in the first half of February. Tomatoes are planted in the developing phase of 7 to 8 leaves and with a visible the first bloom, peppers in the phase of 10 to 12 leaves and a visible branching and the first flower, and cucumbers with 3 to 4 leaves.

The volume of inert substrate per package is most often between 10 and 20 liters. These bags (plates) are 1 m long, 15 to 20 cm wide and 7.5 to 10 cm high. Granular substrates such as perlite and expanded clay can be filled into pots or bags (**Figure 5**). The volume of substrate per plant is most often between 2.5 and 5 liters. Due to the small volume of substrate per plant, frequent fertigation is required, and the number and duration of a single ration depend on the substrate capacity for water (nutrient solution), the development stage of plant microclimatic conditions in the greenhouse.

Planting is done by placing the cube with the seedling/s on the openings provided on the substrate plates. When planting, it is necessary to remove the capillary carrier from the plate and insert it into the cube. Due to favorable temperature and humidity conditions, the rooting lasts for 2 to 3 days and plants continue their growth. Because the substrate is soaked with a nutrient solution before planting, only a few rations of fertigation are needed daily after planting. A few days after planting, the substrate plate is cut into two places, about 2 cm from the bottom of the plate, to allow the runoff of the excess nutrient solution.

2.2.3 Plant care measures and harvest

After rooting, plants are wrapped to prevent the stem breaking. It is necessary to maintain the plants by training them up a vertical supporting twine, removing older leaves as the lower fruit clusters are harvested, and by lowering the main plant stem to keep the whole plant within easy reach of workers.

During vegetation, the daily number and duration of the fertigation rations gradually increases. It is needed to control pH and EC values of nutrient solution in root zone, and ant to perform periodic laboratory analysis of nutrient solution composition. At the same time, in a greenhouse it is necessary to maintain the microclimatic conditions as close as possible to the optimal ones. Harvesting of fruit vegetables in soilless culture is performed at technological maturity, and begins 70 to 90 days after planting for tomato (**Figure 6**) and pepper, and about 30 to 40 days for cucumber. Pepper fruits could be also harvested at physiological maturity. The frequency of harvest depends on the time of harvest and species grown: every 2 to 3 days in cucumbers, every 3 to 5 days in tomatoes and every 10 to 14 days in peppers.

Forty to fifty days before the planned end of the harvest, plants are topped to improve the maturation of formed fruits. A few days before the harvest end, fertigation is stopped.

After the harvest end, plant residues, substrate and parts of the drip irrigation system are moved out from the greenhouse, the greenhouse is cleaned and disinfected

Figure 7.
NFT channel with tomato plants.

and preparations for the next season begin. If the substrate is planned to be reused, it should be stored in a greenhouse to prevent freezing and disrupting the structure.

3. Water culture

Hydroponic techniques for growing plants in a closed system in a nutrient solution without substrate (water culture) are appropriate for growing crops of shorter vegetation, such as leafy vegetables (lettuce, arugula, lamb's lettuce, spinach, Swiss chard, chicory, endive and cress salad) and herbs (parsley, basil, oregano, marjoram, thyme, sage and dill). As the most commonly used in growing leafy vegetables, nutrient film technique, floating hydroponics, ebb and flow and aeroponics could be pointed out.

3.1 Nutrient film technique (NFT)

The nutrient film technique is based on maintaining a thin layer (up to 1 cm) of aerated nutrient solution that continuously flows over the plants root in shallow channels laid under a slope from 0.3 to 2%, which allows the solution to be circulated with a free fall (**Figure 7**). As stated by Velasquez-Gonzales et al. [2], nutrient solution flow can be periodic also. The nutrient solution is supplied by the pump from the container to the channel with the plants, and the solution not used by the plants is collected in the storage tank, analyzed and returned to the system. It is precisely the recirculation of the nutrient solution that is the main advantage of this hydroponic technique. Depending on the culture grown, the channel width varies from 10 to 20 cm, while the maximum length is 20 m.

The channels can be located on the ground or gutters, and are most often made of polymer materials (polyethylene, polyvinyl chloride). The channels contain openings in which seedlings or pots with plants are placed and their root is continuously supplied with water and nutrients, with an ideal solution flow rate of 3 to 8 L/ m^2 per hour for crops such as chrysanthemums and salad. The disadvantages of

Figure 8.
A-frame aeroponics.

this technique are the risk of interrupting the flow of a nutrient solution that very quickly causes root drying, stress and excessive channel warming in the summer due to which young plants may suffer in the initial growth phase. Contrary to growing on substrates, the ion concentration in the root zone does not increase due to continuous solution flow [11].

3.2 Aeroponics

3.2.1 System work out

In aeroponics, the plant root is in the air of dark space, and the nutrient solution is supplied by spraying every 3 to 4 minutes for 15 to 20 seconds in the form of an aerosol, which ensures high humidity (> 95%) in the root zone [11]. The optimum EC and pH values of nutrient solution in aeroponics system lie between 1.5 to 2.5 dS/m and 5.5 to 7.0, sprayed in different intervals, depending on species grown. Nutrient-rich solution is used as a growing medium and provides essential nutrient for sustain plant growth [12]. Velasquez-Gonzales et al. [2] pointed that there is no need for aeration system as oxygen is delivered to root with the sprayed nutrient solution.

In aeroponics, Styrofoam plates with plants are attached to a structure that can be horizontal, or at an angle of 45 to 60 degrees (A-frames). The pump distributes a nutrient solution from the tank to the spray pipe, which is located inside the structure and supply the root of the plant. The nutrient solution is returned to the tank by free fall [11]. The nursery plants might be either raised as seedlings using specially designed lattice pots or cuttings could be placed directly into the system for rapid root formation. Lattice pots allow the root system to develop down into the growth chamber where it is regularly misted with nutrient under controlled conditions [12].

Figure 9.
Stinging nettle in ebb and flow system.

3.2.2 Advantages and disadvantages

The aeroponics provides numerous advantages including a free extension of the root system, direct and sufficient oxygen uptake, and rapid and provision of uniform nutrient spray mist with best root growth environment. Aeroponics uses less water and nutrients because the plant roots are sprayed at intervals using a precise droplet size that could utilize most efficiently by osmosis to nourish the plant [12]. Using A-frames aeroponics (**Figure 8**) results in good utilization of the greenhouse volume because the number of plants has doubled, but due to the variation of light intensity, uneven plant growth may occur.

High initial investments and the application of complex electronic devices justify the application of this hydroponic technique only to high-income cultures [11]. Lakhiar et al. [12] stated that the main problem in aeroponics is related to water nutrient droplet size. The larger droplets permit the less supply of the oxygen availability in the root zone, while the smaller droplets produce too much root hair without developing a lateral root system for sustainable growth. The main potential challenge and drawback of the system is constant power supply throughout the plant growth. Any prolonged rupture of power energy shuts down the nutrient supply and contributes to permanent plant damage.

3.3 Ebb and flow

3.3.1 System workout

The ebb and flow technique is also called "flood and drain" because of its principle of time intervals between dry and wet periods. Nutrient solution is available periodically by soaking the benches filled with plants in containers (**Figure 9**), or with pot plants. After a certain time interval that is programed according to plant species and development stage, the nutrient solution is drained from the bench. The system is closed and the solution is recycled [11, 13].

Benches are covered with an impermeable rigid plastic profile that directs all water to the lowest point at one end of the bench where a siphon device (unpowered) drains nutrient water from the bench surface to a gutter below to return the water to the nutrient storage tank. The supply water is pumped from the water and nutrient management storage tank to each bench or group of benches, filling to a depth of 1–2 cm within 5 min and draining within 10 min for a total water cycle per bay of 15 min. Water and nutrient management system includes freshwater filter and disinfection, nutrient dosing device, storage tank with pump, sensors and controls to distribute irrigation water and nutrients. Mechanical filtering devices are required to remove particulates from the drainage water [14].

3.3.2 Advantages and disadvantages

Ebb and flow has many advantages such as root moisture optimization, water saving and fertilizer saving as compared to top sprinkler irrigation. The nutrient solution concentration may be reduced by up to 50% when compared to nutrient solutions for top sprinkle irrigation, with no detrimental effects on plant growth and quality. Subirrigation systems improve the uniformity and quality of bell pepper and tomato if grown with minimal nutrient and drought stress. When used for potted plants grown on concrete floor, some specific advantages of ebb and flow include:

Figure 10.
Seed sown in Styrofoam plates filled with perlite.

elimination of manual watering, flexibility in design of internal transport of potted plants, heating the root zone with low temperature water, and reducing bacterial and fungal diseases because of cultivation surfaces that were easy to clean and disinfect between cultivation cycles [14].

3.4 Floating hydroponics

The floating hydroponics was first applied in the production of tobacco seedlings, and today they are used efficiently in the production of vegetable seedlings and in the cultivation of leafy vegetables and herbs. It is important to emphasize that in

Figure 11.
Floating hydroponics.

the cultivation of seedlings, the solution is not aerated to prevent the root growth of plants outside the container pot [4].

In floating hydroponics plants are grown in a nutrient solution. The basic advantage of this system is that plants provide access to water, and macro- and micronutrients in the form of ions and oxygen over 24 hours, which they can optimally use during all stages of growth. This results in faster growth and earlier harvesting, which provides more production cycles throughout the year and higher yields [11, 13].

This hydroponic system consists of shallow pools filled with a nutrient solution on which Styrofoam plates or containers with plants float. The nutrient solution is raised capillary through the openings of the pot of containers or the slit of the plates to the substrate in them, that is, to the root of the plant. Styrofoam containers can have a different number of pots, and the plates can be of different dimensions, depending on the type of vegetables and the purpose of cultivation, respectively, whether leafy vegetables are grown due to young leaves for cutting or due to rosette or head. Container pots or slots on plates are filled with perlite or some other substrate into which the seeds of vegetables or herbs are sown (**Figure 10**).

3.4.1 Greenhouse preparation and pool construction

The most demanding part of the work in floating hydroponics growing is preparing the terrain for pool construction, and includes precise straightening, with minimal drop along the greenhouse to keep the water level in all parts of the pool uniform. To allow a simpler pool emptying, it is sufficient to ensure a pool drop of 0.1%. If the surface of the terrain is rough, it is recommended to apply the sand in a layer of 2.5 to 5 cm before rolling and final straightening. Due to the good drainage under the pool, the level of terrain subjected for floating hydroponics construction should be raised 10 to 15 cm above the level of the surrounding terrain. The production surface of the

Figure 12.
Growing lettuce in floating hydroponics.

pool, that is, its width and length, depends on the dimensions of the greenhouse and floating Styrofoam plates or containers for growing leafy vegetables. It is very important that the surface of the entire pool is completely covered with Styrofoam plates to prevent the development of algae that cannot develop without light, and which pollute the nutrient solution and create unfavorable conditions for growing vegetables. The pool frame height should ensure a nutrient solution depth of 20–25 cm and the floating of Styrofoam plates with plants (**Figure 11**).

Agrotextile is first laid on the aligned soil, followed by PE-film of 0.5 mm thick, with complete frame coverage. At the pool bottom, a pipe system for occasional replenishment and daily circulation of the nutrient solution (to enrich the solution with oxygen) is placed. The nutrient solution is gradually added to the pool depending on its consumption, and the transpiration of the plants, respectively. For the entire production of leafy vegetables, it is also recommended to set up a pipe system to maintain the required nutrient temperature [4].

3.4.2 Growing technology

Leafy vegetables (**Figure 12**) harvested by cutting in the developed phase of 5 to 6 leaves (baby leaf) sown in Styrofoam plates (96 × 60 × 2.7 cm), with narrow conical slits filled with perlite of coarse granulation (0 to 6 mm). Sown plates are covered with finer perlite, moistened with water and stacked on each other until seed germination, when the plates are laid in pools filled with aerated nutrient solution. Optimal conditions for germination (temperature from 18 to 20°C and relative humidity around 95%) are provided in the germination chamber [4].

If leafy vegetables is grown for harvest of rosettes or heads, seeds are sown into rockwool plugs (cubes) 3 × 3 cm. Cubes with seedlings are placed in lattice pots, in holes (planting sites) distanced 20 × 20 cm in Styrofoam plates [11]. The plates with seedlings are laid in pools filled with a nutrient solution of a certain chemical composition and optimal temperature.

In this hydroponic technique, plants are constantly absorbing a nutrient solution, especially at higher air temperatures when transpiration is more intense, so the level of the solution decreases and it is necessary to ensure a pool supplement. The pH and EC values, the amount of dissolved oxygen and the nutrient solution temperature should be measured daily, and the nutrient solution composition by chemical analysis should be done every 2 weeks. The optimal pH value of the solution is from 5.8 to 6.2, while the EC value in leafy vegetable cultivation should be in the range between 2.5 (lettuce, lamb's lettuce) and 3.2 dS/m (arugula). The availability of nutrients for plant is affected by the pH value and temperature of the nutrient solution and the amount of dissolved oxygen in the solution. The recommended temperature of the leafy vegetable growing solution should be from 21 to 23°C, while the optimal amount of dissolved oxygen is 4 to 9 mg/L [13, 15]. If the solution temperature is higher, the ability of the solution to retain oxygen decreases and the breathing of the roots is more intense and oxygen consumption is higher. Lack of oxygen in the nutrient solution (below 3 mg/L) results in less root permeability to the water so the plant cannot adopt nutrients in the required amount, and toxin accumulation can occur. Plant growth is slower and plant damage and leaf chlorosis are possible. Lowering the solution temperatures too high will ensure that larger amounts of oxygen are retained and root respiratory is reduced [16].

The length of the vegetation from sowing to harvest depends on the type of leafy vegetables and growing conditions, and the equipment of the protected area (side and roof ventilation, heating and shading equipment and supplemental lighting, nutrient solution heating and cooling system). Lettuce, lamb's lettuce, endive and chicory are harvested once, while arugula and herbs can be harvested repeatedly. However, the vegetation tip should not be damaged during the first harvest, so the plants could grow again. The annual yield of arugula and lamb's lettuce in floating hydroponics may be 40 to 50% higher than the yield in the case of soil grown in greenhouse [17, 18].

After a year-round production period, the pools are cleaned from the rest of the nutrient solution, perlite particles and organic matter, than disinfected and prepared for a new year-round cycle with the preparation of a nutrient solution, filling the pool and continuous sowing and harvesting.

4. Water and nutrient solution

4.1 Water quality

Water is the basis of any nutrient solution and therefore, it is necessary to provide sufficient amounts of quality water. High water quality is determined by the low concentration of dissolved substances, especially salts. The higher the water quality, the easier it is for producers to formulate an optimal nutrient solution. If the water quality is lower, more water is needed to dissolve the nutrient salts in open systems, that is, to remove excess salt from closed systems. Low quality can be supplemented by more water [4].

The quality of water should be taken into account at each beginning of the production season in the greenhouse since low-quality water is not usable and is expensive to "process" by filtration and/or reverse osmosis. Quality primarily depends on the available water source (rainwater, surface water-treated waste water and ground water). Rainwater is one of the best sources regarding quality. Before water can be used, it must be analyzed to determine the basic level of all minerals and ions (Ca^{2+}, Mg^{2+}, SO_4^{2-}, HCO_3^-, Na^+, Cl^-,) present and the pH and alkalinity. Without this information, it will be difficult to prepare the optimal nutrient solution [19, 20]. Water quality depends on the concentration of dissolved substances, and the presence of microorganisms such as algae, fungi and bacteria, and certain sediments. The overall analysis should show anions and cations, and special attention should be paid to salinity, alkalinity, and excessive concentrations of sodium, sulfate, and chloride. When using a drip irrigation system, high water quality is required to avoid possible interference by clogging the droppers with iron and manganese [20].

The required amount of water is mainly determined by microclimatic conditions and a leaf surface [4], which also affects the optimal composition of a nutrient solution [6] and EC value [19]. Under conditions of high humidity, low light and low temperature, water consumption can be very low. It is very important to know how to estimate the maximum amount of water used when the irrigation system is constructed and installed. The amount of water that plants consume is caused by the degree of growth of the plant, solar radiation, relative humidity and air movement.

Salinity is the amount of all dissolved salts quantified as water electrical conductivity (EC value), and is expressed in mS/cm or dS/m. An important assumption is that the EC value of the spring water should be below 1 dS/m. In some cases, the use of water with higher EC value is possible for so long while ions, which cause high

Figure 13.
Fertigation unit.

EC-value, are used as plant nutrients. Even then, the concentration of these ions should not be excessive. Water with EC value 0.75–2.25 dS/m has slight to moderate restriction in use, while with >2.25 it has severe restrictions. The use of salted water in hydroponic cultivation in arid areas results in a slightly lower yield of cultivated crops, but therefore of excellent quality. Harmful effects on plant growth are caused by water and salinity stress. The maximum acceptable level of Na^+ in the nutrient solution varies between 1 and 8 mmol/l, while the maximum acceptable Cl^- level in the root zone is 0.2–0.5 mmol/l higher than the maximum acceptable level for Na^+ [19–21].

4.2 Nutrient solution preparation and distribution

In addition to water, nutrient salts or water-soluble complex fertilizers and acids are necessary to prepare a nutrient solution. The advantage of nutrient salts is that they represent high-purity chemical compounds composed of two to three nutrients. Complex water-soluble fertilizers most often contain nitrogen phosphorus, potassium and magnesium with the addition of microelements, which means that when correcting the composition of the nutrient solution, it is not possible to change the concentration of only one nutrient than to change all the concentrations of all nutrients found in the fertilizer [4]. Acid (nitric or phosphoric) needs to be added to the nutrient solution to lower the pH value of water (7.2 to 7.5) to optimal for hydroponic cultivation, which is between 5.5 and 6.8 [8, 17, 19, 21], although values between 5.0–5.5 and 6.5–7.0 may not cause problems in most crops [22, 23]. The EC value measured in fresh nutrient solution ranges from 1.5 to 3 dS/m [1, 8]. Lieth and Oki [24] stated that EC in soilless production may vary between 0 and 5 dS/m. It has been advised to maintain the EC below 3 dS/m to assure rapid plant growth, but this is impossible if the water is high in dissolved salts, and the addition of nutrients will raise EC to higher value than 3 dS/m.

	Tomato	Pepper	Cucumber	Lettuce*	Strawberry	Seedlings
Macronutrients, mmol/L						
NO_3^-	13.75–16.00	15.50–16.00	16.00	16.00–19.00	11.25–12.00	15.00–16.75
$H_2PO_4^-$	1.25–1.50	1.25–1.75	1.25–1.50	1.50–2.00	1.00–1.25	1.50–2.50
SO_4^{2-}	3.75–4.40	1.75	1.375–1.50	1.125–2.00	1.50	2.50–3.00
NH_4^+	1.20–1.25	0.75–1.25	1.25	1.00–1.25	1.00	1.25–2.00
K^+	8.75–9.50	6.50–7.00	8.00	9.50–11.00	4.80–5.50	6.00–8.00
Ca^{2+}	4.25–5.40	5.00	4.00	4.50	3.50–3.60	4.00–5.00
Mg^{2+}	2.00–2.40	1.50	1.375–1.50	1.00	1.35–1.50	3.00–3.50
Micronutrients, μmol/L						
Fe^{3+}	15.00	15.00	15.00	40.00	20.00–30.00	25.00
Mn^{2+}	10.00	10.00	10.00	5.00–7.00	10.00	10.00–15.00
B^{3+}	30.00	30.00–35.00	25.00	30.00–40.00	10.00–15.00	35.00
Zn^{2+}	5.00	5.00	5.00	4.00–7.00	7.00	5.00
Cu^{2+}	0.75	0.75–1.00	0.75	0.75–1.00	0.75	1.00
Mo^{6+}	0.50	0.50	0.50	0.50–1.00	0.50	0.50
EC, dS/m	2.30–2.60	2.20	2.20	2.20–3.20	1.60	2.20–2.60
pH	5.5–6.2	5.5–6.2	5.5–6.2	5.8–6.2	5.5–6.2	5.5–6.2

Lettuce and leafy vegetables.

Table 2.
Nutrient solution composition in tank for greenhouse crops according to different authors [9, 19, 21–23].

The preparation of fresh nutrient solution is performed using a dosatron, mixer or fertigation unit (**Figure 13**) depending on the greenhouse. Regardless of the hydroponic cultivation technique, the finished nutrient solution is prepared from 100-fold concentrated solutions in relation to the concentration of the solution that is brought to plants by the system. Therefore, in each hydroponic production there are at least three tanks for concentrated solution [11]. Two tanks are filled with different stock solutions to separate calcium from sulfate and phosphate fertilizers, thereby avoiding precipitation of low-soluble compounds. The third tank contains a solution of nitric or phosphoric acid, which serves to regulate the solution pH value, by neutralization of HCO_3^- ion [19]. An equal volume of stock solutions used for fresh solution preparation is necessary in order to avoid nutrient misbalance. The volume of acid used depends on the water pH and the desired pH value of the nutrient solution [4].

In modern hydroponic growing systems, the nutrient solution parameters (oxygen concentration, temperature, pH and EC) are automatically controlled by a computer system that uses special sensors. The software sets the target values, and the fertigation unit measures water parameters and compares them with target values to add proper volumes of concentrated solutions and acid until the target values are reached. Additionally, probes are immersed in the growing substrate or in nutrient solution to collect data in root zone. The data is transmitted in real time to the cloud, from where it can be read at any time *via* a mobile app or computer. In this way, a faster response is possible when the parameters of the nutrient solution need to be corrected, which undoubtedly has a positive effect on the success of the cultivation [25].

Lieth and Oki [24] stated that nutrient solution in hydroponic growing systems could be delivered to plants by overhead, surface or subsurface irrigation. However, the dominant way of irrigation is surface, particularly drip irrigation in substrate grown crops. Nutrient solution is delivered by drippers, pinned or laid on the upper side of substrate. One of the most significant problems of drip irrigation is dripper clogging, mechanically or chemically, directly related to the quality of irrigation water and its physical, chemical and microbiological properties. Therefore, a water quality analysis should be performed before installing the drip irrigation system. The filtering site must certainly be an integral part of the drip irrigation system [4].

4.3 Nutrient solution composition

Although there are no significant differences in the nutrient solution composition among crops, crops can vary significantly in the absorption of individual nutrients, especially in certain parts of the vegetation. Nutrient absorption is affected by many abiotic (air and substrate temperature and humidity, light intensity and CO_2 concentration) and biotic (growth and development phase, fruit load and pest presence) factors. However, for more or less standardized growing conditions on substrates, a strong correlation between fresh fruit yield and nutrient absorption has been established [4, 23]. Contrarily, Sonneveld and Voogt [19] and Savvas et al. [26] quote that specialized nutrient solution for each greenhouse crop or even for developmental stage is available. Use of this kind of solution is optimal when nutrient uptake ratios are similar with the relative proportions between the same nutrients in fresh solution. This principle should be strictly followed in closed hydroponic systems to avoid

Figure 14.
UV-sterilization unit.

nutrient accumulation and/or depletion. Vox et al. [27] stated that the more concentrated nutrient solutions are used for fast-growing crops, such as vegetables, while for ornamental plants and strawberry lower nutrient concentrations are normally used. Plenty of different nutrient solution formulas have been published and some of them are summarized in **Table 2**.

4.4 Nutrient solution sterilization and recirculation

According to the use of a nutrient solution, hydroponic systems are divided into the following: open ones where once used nutrient solution is not used again in the system but is drained into evaporation channels or used to fertilize soil-produced crop; and closed ones where drained nutrient solution is passed through a sterilization system, supplemented with a fresh nutrient solution and reused [6]. If the hydroponic system is open, the irrigation system should ensure the amount of nutrient solution or water, which will maintain or reduce the salt concentration. Due to that, the part of supplied nutrient solution should be drained from the substrate. In practice, the drained solution volume varies between 10 and 30%, depending on the quality of the water and/or on the crop sensitivity to salinity [4, 6, 23, 27]. In closed systems, salt accumulation in the root zone is more common, resulting in reduced yields. To avoid this kind of problem, the nutrient concentrations and injection rates of fresh and recycled nutrient solution should be monitored and regulated. Also, irrigation with freshwater, which washes away excess nutrients, could be applied.

Root's zone in hydroponic systems needs to be pathogen free to efficiently produce good-quality products [2]. Due to hydration, there is a high potential for the rapid spread of root diseases [28], especially in closed hydroponic systems. Closed hydroponic systems reduce or limit the runoff of drained nutrient solution into the environment [3], so in closed systems the drained solution should be filtered and disinfected before it is recycled, to avoid spread of pathogens [29]. There are five main methods of pathogen control in these systems: heat, filtration, chemical, radiation and biological control. Sterilization (heat, oxidizing chemicals and UV-radiation) and membrane filtration methods are generally very effective, but may adversely affect beneficial microorganisms in the recirculated solution (**Figure 14**). Slow filtration and microbial inoculation methods are less disruptive of the microflora, but effectiveness may vary with the pathogen. Microbial inoculation is perspective in targeted disease suppression, but still just a few products are commercially available [28, 29].

From a sustainability perspective, it is important to recirculate the nutrient solution to minimize water consumption and residuals to dispose into environment. However, it is not always possible to implement systems that balance the consumption of natural resources, energy and financial costs [2]. Besides the environmental benefits, closed hydroponic systems can provide higher economic profits, since they reduce the quantity of water and fertilizers used during production, and they are more efficient in using water and nutrients than open systems, respectively [30]. In their research, De la Rosa-Rodríguez et al. [30] achieved 26.9% (13.5 kg) higher tomato yield per liter of water in closed than in the open system.

5. Conclusions

Hydroponic growing systems include plant growing techniques without soil, on inert substrate (soilless culture), or without substrate (water culture). Inert substrates used are mainly of inorganic or organic origin. The advantage of organic substrate use is their sustainability with no or minimal impact to the environment, so they could be recommended. Water culture techniques represent closed hydroponic systems, which are more efficient in water and fertilizer use compared to open systems (mostly on substrates), and especially compared to soil production. Due to high-quality yield regardless of grown crop, hydroponic systems could be a way to increase the food production sustainability in the future, characterized by population growth, climate changes and the reduction of natural resources.

Future development of hydroponics through research and particularly through application should be focused on vertical farming and plant factories, which will ensure continuous production increase with sustainable use of resources by controlled environment agriculture.

Conflict of interest

The authors declare no conflict of interest.

Author details

Božidar Benko*, Sanja Fabek Uher, Sanja Radman and Nevena Opačić
Faculty of Agriculture, University of Zagreb, Zagreb, Croatia

*Address all correspondence to: bbenko@agr.hr

References

[1] Gruda N. Soilless culture systems and growing media in horticulture: An overview. In: Gruda N, editor. Advances in Horticultural Soilless Culture. 1st ed. Cambridge, UK: Burleigh Dodds Science Publishing Limited; 2021. pp. 1-20

[2] Velazquez-Gonzalez RS, Garcia-Garcia AL, Ventura-Zapata E, Barceinas-Sanchez JDO, Sosa-Savedra JC. A review on hydroponics and the technologies associated for medium- and small-scale operations. Agriculture. 2022;**12**:646. DOI: 10.3390/agriculture12050646

[3] Ampim PAY, Obeng E, Olvera-Gonzalez E. Indoor Vegetable Production: An Alternative Approach to Increasing Cultivation. Plants. 2022;**11**:2843. DOI: 10.3390/plants11212843

[4] Borošić J, Benko B, Toth N. Soilless Culture of Vegetables (in Croatian). Croatia: Faculty of Agriculture, University of Zagreb; 2011. 126 p

[5] van Delden SH, Sharath Kumar M, Butturini M, Graamans LJA, Heuvelink E, Kacira M, et al. Current status and future challenges in implementing and upscaling vertical farming systems. Nature Food. 2021;**2**:944-956. DOI: 10.1038/s43016-021-00402-w

[6] Savvas D, Gruda N. Application of soilless culture technologies in the modern greenhouse industry—A review. European Journal of Horticultural Science. 2018;**83**(5):280-293. DOI: 10.17660/eJHS.2018/83.5.2

[7] Gitnux Marketdata: The Most Surprising Hydroponics Industry Statistics and Trends in 2023. [Internet] 2023. Available from: https://blog.gitnux.com/hydroponics-industry-statistics/ [Accessed: August 03, 2023]

[8] Popsimonova G, Benko B, Karic L, Gruda N. Production systems in southeastern European greenhouses. Acta Horticulturae. 2021;**1326**:137-144. DOI: 10.17660/ActaHortic.2021.1326.18

[9] Enzo M, Gianquinto G, Lazzarin R, Pimpini F, Sambo P. Principi tecnico-agronomici della fertirrigazione e del fuori suolo. Padova, Italy: Tipografia-Garbin; 2001. 204 p

[10] Raviv M, Wallach R, Silber A, Bar-Tal A. Substrates and their analysis. In: Savvas D, Passam H, editors. Hydroponics Production of Vegetables and Ornamentals. Athens, Greece: Embryo Publications; 2002. pp. 25-101

[11] Opačić N, Radman S, Fabek Uher S, Benko B, Voća S, Šic Žlabur J. Nettle cultivation practices—From open field to modern hydroponics: A case study of specialized metabolites. Plants. 2022;**11**:483. DOI: 10.3390/plants11040483

[12] Lakhiar IA, Gao J, Syed TN, Chandio FA, Buttar NA. Modern plant cultivation technologies in agriculture under controlled environment: A review on aeroponics. Journal of Plant Interactions. 2018;**13**(1):338-352. DOI: 10.1080/17429145.2018.1472308

[13] Nicola S, Hoeberechts J, Fontana E. Ebb-and-flow and floating systems to grow leafy vegetables: A review for rocket, corn salad, garden cress and purslane. Acta Horticulturae. 2006;**747**:585-593

[14] Giacomelli GA, van Weel PA, Blok C. Ebb and flood nutrient delivery system for sustainable automated crop production. Acta Horticulturae.

2020;**1296**:1129-1136. DOI: 10.17660/ActaHortic.2020.1296.143

[15] Conesa E, Fernández JA, Niñirola D, Egea-Gilabert C. Nutrient solution aeration and growing cycles affect quality and yield of fresh-cut baby leaf red lettuce. Agricultural and Food Science. 2015;**24**(4):313-322. DOI: 10.23986/afsci.52792

[16] Karnoutsos P, Karagiovanidis M, Bantis F, Chatzistathis T, Koukounaras A, Ntinas GK. Controlled root-zone temperature effect on baby leaf vegetables yield and quality in a floating system under mild and extreme weather conditions. Journal of the Science of Food and Agriculture. 2021;**101**:3933-3941. DOI: 10.1002/jsfa.11033

[17] Toth N, Fabek S, Benko B, Žutić I, Stubljar S, Zeher S. The effect of abiotic factors, sowing density and multiple harvest to arugula yield in floating hydroponic. Glasnik Zaštite Bilja. 2012;**35**(5):24-34

[18] Fabek S, Toth N, Benko B, Borošić J, Zutić I, Novak B. Lamb's lettuce growing cycle and yield as affected by abiotic factors. Acta Horticulturae. 2011;**893**:887-894. DOI: 10.17660/ActaHortic.2011.893.98

[19] Sonneveld C, Voogt W. Plant Nutrition of Greenhouse Crops. Dordrecht: Springer; 2009; 431 p. DOI: 10.1007/978-90-481-2532-6

[20] van Os E, Blok C, Voogt W, Waked L. Water Quality and Salinity Aspects in Hydroponic Cultivation [Internet]. Netherlands: Wageningen University and Research; 2016. Available from: https://edepot.wur.nl/403810 [Accessed: June 15, 2023]

[21] Nutrient solutions for greenhouse crops [Internet]. 2016. Available from: https://cdnmedia.eurofins.com/corporate-eurofins/media/12142795/160825_manual_nutrient_solutions_digital_en.pdf [Accessed: June 15, 2023]

[22] Adams P. Nutritional control in hydroponics. In: Savvas D, Passam H, editors. Hydroponic Production of Vegetables and Ornamentals. Athens, Greece: Embryo Publications; 2002. pp. 211-261

[23] Silber A, Bar-Tal A. Nutrition of substrate-grown plants. In: Raviv M, Lieth JH, editors. Soilless Culture: Theory and Practice. USA: Elsevier; 2008. pp. 291-339

[24] Lieth JH, Oki LR. Irrigation in soilless production. In: Raviv M, Lieth JH, editors. Soilless Culture: Theory and Practice. Amsterdam, Netherlands: Elsevier; 2008. pp. 117-156. DOI: 10.1016/B978-044452975-6.50006-X

[25] Cafuta D, Dodig I, Cesar I, Kramberger T. Developing a modern greenhouse scientific research facility—A case study. Sensors. 2021;**21**(8):2575. DOI: 10.3390/s21082575

[26] Savvas D, Gianquinto GP, Tüzel Y, Gruda N. Soilless culture. In: Good Agricultural Practices for Greenhouse Vegetable Crops—Principles for Mediterranean Climate Areas. Rome, Italy: Plant Production and Protection Paper 217, FAO; 2013. pp. 303-354

[27] Vox G, Teitel M, Pardossi A, Minuto A, Tinivella F, Schettini E. Sustainable greenhouse systems. In: Salazar A, Rios I, editors. Sustainable Agriculture: Technology, Planning and Management. New York, USA: Nova Publishers; 2010. pp. 1-79

[28] Ehret DL, Alsanius B, Wohanka W, Menzies JG, Utkhede R. Disinfestation of recirculating nutrient solutions in greenhouse horticulture. Agronomie.

2001;**21**(4):323-339. DOI: 10.1051/agro:2001127

[29] Wohanka W. Nutrient solution disinfection. In: Savvas D, Passam H, editors. Hydroponic Production of Vegetables and Ornamentals. Athens, Greece: Embryo Publications; 2002. pp. 345-372

[30] de la Rosa-Rodríguez R, Lara-Herrera A, Trejo-Téllez LI, Padilla-Bernal LE, Solis-Sánchez LO, Ortiz-Rodríguez JM. Water and fertilizers use efficiency in two hydroponic systems for tomato production. Horticultura Brasileira. 2020;**38**:47-52. DOI: 10.1590/S0102-053620200107

Chapter 3

Agriculture's Contribution to the Emission of Greenhouse Gas Nitrous Oxide (N_2O) and Its Feasible Mitigation Strategies

Raushan Kumar and Nirmali Bordoloi

Abstract

Climate change and agriculture have a dual mode of relationship. Agriculture is an important sector of the country's economy and it significantly contributes to climate change by releasing greenhouse gases (GHGs) to the atmosphere. On the other hand, climate change is a global threat to food security and it can affect agriculture through variation of weather parameters. Reducing GHGs emission mainly methane (CH_4) and nitrous oxide (N_2O) from the agriculture could play a significant role in climate change mitigation. N_2O is a potent greenhouse gas mainly emitted from rice-wheat cropping system. Agricultural lands are considered as one of the important anthropogenic sources of N_2O emissions and it account almost 69% of the annual atmospheric N_2O emission and application of commercial fertilizers is considered as a major contributor to the N_2O emission. This book chapter focuses on the feasible soil and crop management practices to reduce the N_2O emission from agriculture without compromising the productivity. Different environmental factors that have a major impact on N_2O production are also discussed in this chapter. On urgent basis, the world needs to reduce the anthropogenic N_2O emissions from agriculture and adapt its sustainable cropping system and food-production system to survive with climate change.

Keywords: climate change, food security, fertilizer, nitrous oxide, management practices

1. Introduction

Global climate change is caused by the increasing concentration of many climate pollutants like carbon dioxide (CO_2), methane (CH_4) and nitrous oxide (N_2O) etc. The agriculture and food production is connected with emissions of all these three gases but emissions of CH_4 and N_2O are directly dominated by agricultural activities [1] and 10–12% of the total GHGs produced globally by anthropogenic activities [2]. Among the non-CO_2 greenhouse gases (GHGs), N_2O is an important long lived GHG and agriculture represents its largest source worldwide. N_2O is a major driver of climate change and considered as a very reactive gas and potent ozone-depleting substance in the stratosphere [3]. Moreover, it exerts adverse impacts on crop production and human health [4]. The emission of N_2O can lead to an indirect health impact, namely the depletion of the stratospheric ozone layer. This depletion results in higher levels of UV radiation reaching the earth's surface, leading to an increased incidence of skin cancers [5]. Additionally, regions with elevated N_2O concentrations may experience air pollution due to its contribution. When N_2O combines with other pollutants, it can form ground-level ozone and fine particulate matter, which can worsen respiratory issues, particularly in individuals who already have asthma and chronic obstructive pulmonary disease (COPD) [5]. The rising earth's temperature due to the increasing N_2O concentration can have also detrimental effects on precipitation patterns and lead to more extreme temperatures, adversely impacting plant growth and productivity. Additionally, increased N_2O levels in the atmosphere can cause higher nitrogen deposition in soils. While nitrogen is vital for plant growth but excessive amounts can disrupt the nutrient balance, depleting essential nutrients and compromising plant health [5]. Furthermore, the depletion of the ozone layer due to the emission of N_2O allows harmful UV radiation to reach the earth's surface, potentially harming plants and hindering the process of photosynthesis.

Since 1750, concentrations of GHGs have been increasing due to anthropogenic activities. The anthropogenic N_2O is increasing annually, which has risen from a pre-industrial value of 270 ppb to a value of 324 ppb in 2011 and 332 ppb in 2019 [6].

Agriculture is the major primary anthropogenic source of N_2O emission, globally contributing around 3.8 (2.5–5.8) Tg N yr^{-1} or 22% to the atmospheric N_2O budget [7]. The use of synthetic fertilizer, manure and increase in agricultural lands are the main reason of N_2O emissions from soil (**Figure 1**). When plant roots cannot uptake all the applied fertilizer due to their growth stages, some of it runs off or leached out and remaining amount is consumed by the soil microbes and convert the ammonia to nitrate and finally back to N_2 gas (**Figure 2**).

N_2O is emitted as a byproduct during the conversion of ammonia/ammonium to nitrate and nitrate to N_2 by microbial process of nitrification and denitrification respectively [8]. The excess nitrogen in the soil also leads to lower nitrogen use efficiency (NUE) by plants. Although the global agricultural food system depends of application of synthetic fertilizers to increase the crop productivity however; the

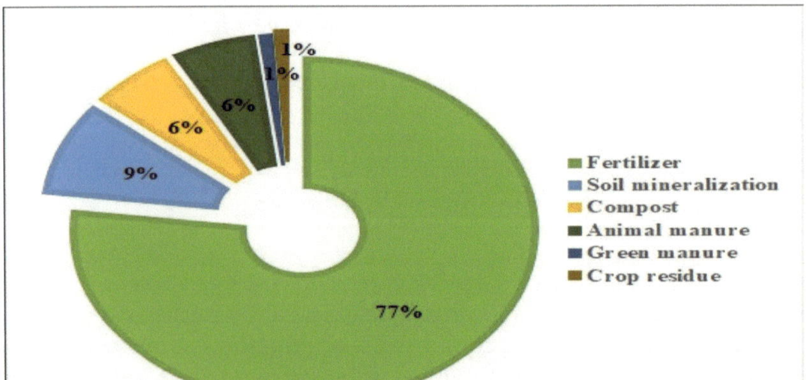

Figure 1.
Contribution of different sources to N_2O emission from soil (source: Gupta et al. [8]).

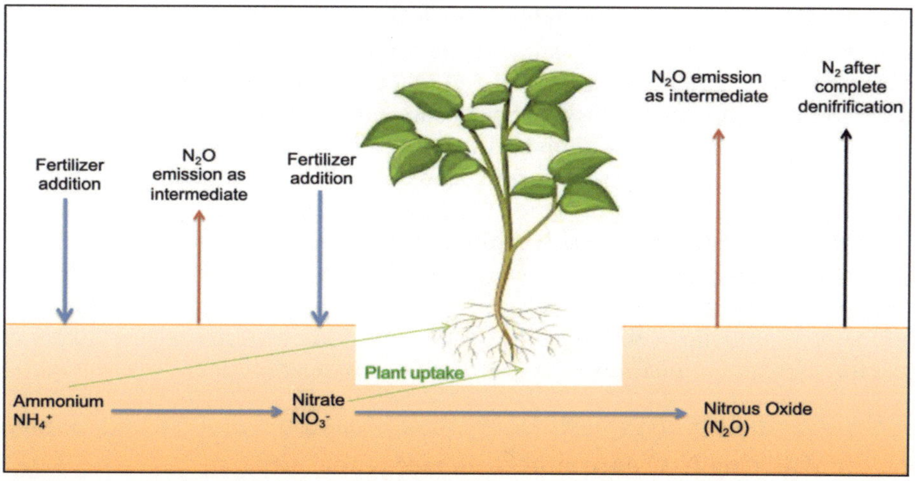

Figure 2.
Use of excess nitrogen and N_2O emission from the soil.

abundance use of synthetic fertilizer is unsustainable due emission of N_2O from soil and pollutes waterways through nitrate leaching. The global food system is responsible for ~21–37% of annual emissions [9]. Further, N_2O emissions are expected to increase over to coming decades due to projected increases in food demand for over increasing population, agricultural land and fertilizer use. However, active management of agroecosystems through managing soil and plants can offer a sustainable opportunity for N_2O mitigation without jeopardizing crop growth and food production. In this chapter, we have tried to address all the factors associated with agricultural N_2O emission and their feasible management practices to reduce the production and emission of N_2O.

2. Role of rice-wheat cultivation in N_2O production and emission

The primary sources of N_2O in rice-wheat soil is the transformation of reactive N by soil microbes [10]. When N enters the soil in the form of NH_4^+ and NO_3^- via organic or mineral fertilizers, various reactions might occur, resulting in N_2O production. Three main processes, namely nitrification, denitrification and nitrifier denitrification, are considered the main contributors to N_2O emissions [11]. Nitrification (NF) is regarded as the primary process involved in the global N cycle. The majority of N transformation during nitrification is mediated by autotrophic microorganisms. The initial stage in NF is NH_3 oxidation to hydroxylamine. This mechanism is mediated by ammonia-oxidizing archaea (AOA) and ammonia-oxidizing bacteria (AOB). Denitrification (DNF) is a reduction process involving the conversion of NO_3 to N_2, mediated by facultative anaerobic bacteria [12]. This process can be completed up to N_2 production, but if it is not completed, N is released as NO and N_2O. 70% of worldwide N_2O emissions are attributed to NF and DNF microbial activities [13]. Nitrifier denitrification is the reduction of NO_2- to NO, then to N_2O and finally to N_2 [14]. The soil gets submerged or saturated with water during rice cultivation. This reduces the amount of oxygen available to nitrifying microorganisms, halting the nitrification process. In such soils NH_4-N is the major form of N. The drying of the soil at the harvest of rice crop and aerobic condition of soil in wheat cultivation favors nitrification and accumulation of NO_3-N, which is prone to losses by denitrification and leaching during flooding in subsequent rice cultivation. Moreover, the fluctuating soil moisture conditions and the intermittent drying and flash flooding in rice cultivation, cause large N losses to occur. Therefore, though continuously flooded rice paddies are not considered to be an important source of atmospheric N_2O because N_2O, an intermediary product of denitrification, would be rapidly reduced to N_2 under the intensive anaerobic conditions and rice-wheat systems may produce considerable amount of N_2O. Each process's contribution to N_2O emission is affected by soil texture, organic C, soil pH, microbial activity, and environmental factors such as precipitation and temperature [15], as discussed in next section.

3. Factor affecting N_2O emission from rice wheat soil

N_2O production and emissions from rice wheat soil are regularly governed by different microbial-mediated activity and also depends on several pathways of gas transport, such as: plant-mediated transport (through the aerenchyma). N_2O emission from the rice wheat soil are also mediated through biologically, therefore, its emission from the soil is affected by different climatic as well as agricultural management factor which are depicted in **Figure 3**.

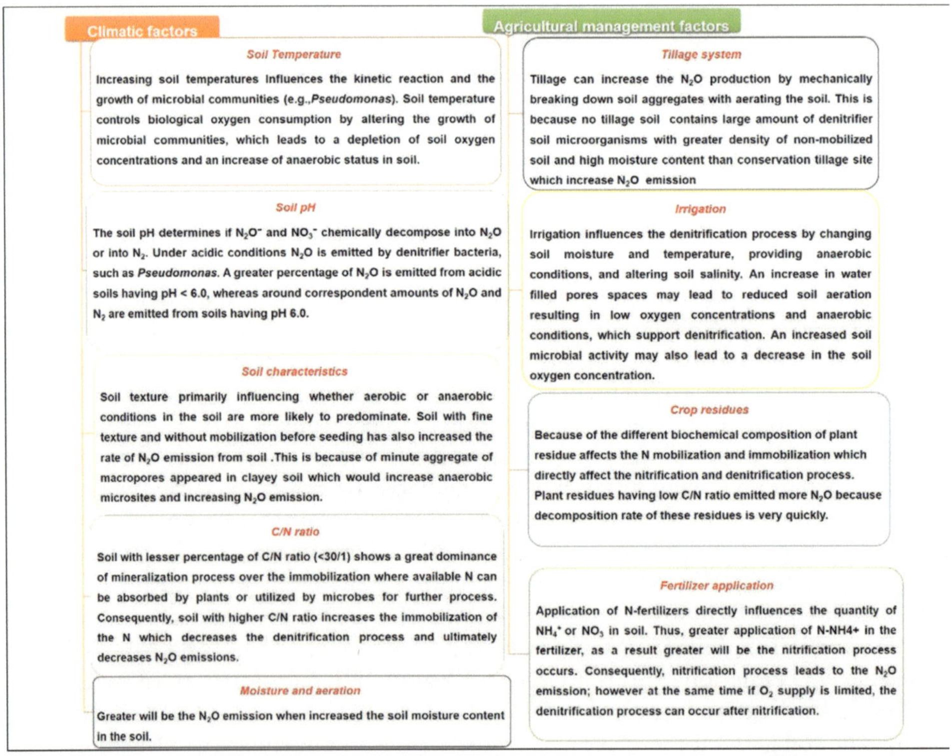

Figure 3.
Factors affecting N_2O emission from rice wheat ecosystem.

4. Sustainable mitigation strategies of N_2O emissions

There are a number of mitigation strategies that can be applied to rice and wheat grown soil that would increase productivity while lowering N_2O emissions and strengthening agriculture's ability to withstand climate change. In this section, briefly we draw attention to some recent research advances in mitigation strategies and technology tools to expand our understanding about soil and crop management for enhanced nitrogen use efficiency (NUE) and N_2O emission mitigation (**Figure 4**). All mitigation strategies focus on site-specific management practices and the use of technologies that will assist limit N losses via ammonia volatilization and nitrate runoff, leaching and drainage pathways. The importance of site-specific agricultural management practices to improve crop and soil recovery of applied N (efficiency), crop productivity per unit of N applied (efficacy), and N_2O per unit of crop production has been stressed.

4.1 Agricultural management practices

4.1.1 Irrigation pattern management

Flood irrigation (FI) is the most widely used irrigation method in developing countries such as India, Pakistan, Bangladesh, and most part of Africa. High volumes of water are given to crops in FI, resulting in fertilizers dilution and easily absorbed [16]. Large irrigation volumes, on the other hand, influence the anaerobic conditions permissive to N_2O generation and nitrate leaching [17]. To avoid this, a precise water application strategy, such as alternate wetting and drying (AWD), could save water while also lowering N_2O emissions. This is because low water content requires more time for oxygen penetration into the soil, which leads to inhibition of microbial activity in the soil responsible for N_2O formation [18]. Similarly, intermittent irrigation, which means the field is alternately watered and drained, has a high potential to reduce N_2O production from soil because this irrigation method has the advantage of improving soil oxidative conditions by increasing root activity, soil bearing capacity, and ultimately minimizing water inputs that create anaerobic conditions. This

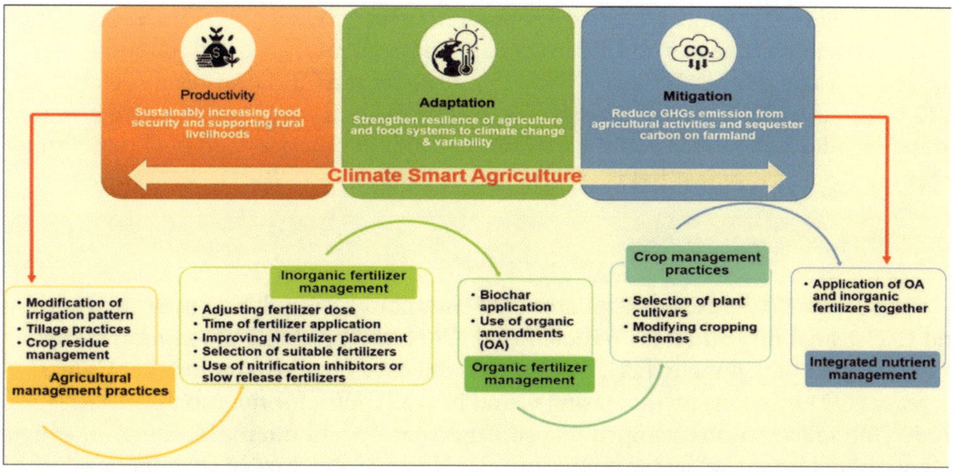

Figure 4.
Key principles of climate smart agriculture and associated mitigation strategies of N_2O emissions.

Crops	Agricultural management practices	N_2O mitigation potential	References
Rice	AWDI + RS 15 t ha^{-1}, AWDI + RS 30 t ha^{-1}	18.68%, 31.55%	[20]
Wheat	Sub-surface drip irrigation	56.16%	[21]
Wheat	Straw incorporation	19.4%	[21]
Rice	Reduce tillage	3–6%	[22]
Rice	Zero tillage	22%	[23]
Rice	Optimizing N rate with RT	6%	[24]
Wheat	Zero tilled with rice residue application	11–12.8%	[25]

Table 1.
Different agricultural management practices and N_2O mitigation potential from rice-wheat fields.

promotes the penetration of oxygen into the paddy soils and, as a result, reduces N_2O emissions. Another modified irrigation strategy is sprinkler-irrigated field (SI), the surface layer in a SI is comparatively loose than FI. As a result, in such soils, the NO_3-N and NH_4-N are less leached and remain more concentrated in the root zone, making them more easily absorbed by plant roots and hence less likely to be converted to N_2O [19]. Different irrigation pattern and N_2O mitigation potential from rice-wheat fields are showed in **Table 1**.

4.1.2 Tillage practices

Soil tillage has a significant impact on N_2O emissions during rice-wheat cultivation because it alters soil physiochemical and biological characteristics, stimulating microbial N_2O generation [26]. Traditional plowing or rotational tillage, which is extensively employed today, exposes the surface, which increases soil depletion and lowers the quality of cultivated land as well as the soil's ability to continually feed fertilizer. The usage of conservation tillage (CT) techniques, such as no-tillage (NT) and reduced tillage (RT), is progressively increasing, owing to the reduction of greenhouse gases, improvement of soil and water quality and enhanced water efficiency. Six et al. [27] proposed that preserving NT throughout time could lower N_2O emissions. These findings are also corroborated by van Kessel [28]. The researchers conducted a meta-analysis on 239 direct comparisons of CT, NT and RT and found that, on average, neither NT nor RT emit more N_2O than CT. Long-term research (>10 years) using NT and RT procedures, primarily in dry regions, revealed a considerable reduction in N_2O emissions. Different tillage practices and N_2O mitigation potential from rice-wheat fields are showed in **Table 1**.

4.1.3 Crop residue management

Crop residue (CR) return regulates N_2O emissions by regulating microbial activity and C/N availability and it is predicted that CR return produces 0.4 million metric tons of N_2O-N yr^{-1} globally [29]. Several authors have noted that returning CR can increase N_2O emissions by increasing C and N availability for microbial activities and modifying soil aeration by improving soil aggregation and microbial demand, which is thought to be a major factor mediating soil NF and DNF for N_2O production [29]. Other authors, on the other hand, reported that adding CR had an inhibitory effect on N_2O emission, depending on soil conditions and crop residue C/N ratio [30]. The return of CR can act as a carbon source for microbial development, promoting N uptake by microorganisms. This activity can result in a fierce competition for NH_4+ between heterotrophic microorganisms and autotrophic nitrifiers, which results in N_2O production [31]. However, in CR management, it is believed that no unambiguous behavior with regard to N_2O emission can be detected. To improve smart CR management and its contribution to reduced N_2O emissions, several factors must be considered, including CR properties and ambient circumstances.

4.2 Inorganic fertilizer management

Mitigating N_2O emissions requires increased NUE through improved temporal synchrony between N supply and plant demand. This requires efficient N management strategies, such as selection of the right source (enhanced efficiency fertilizers), right quantity, right time and right application method.

4.2.1 Altering fertilizer dose and matching N supply with crop demand

Appropriate fertilizer management can significantly reduce N_2O emissions from rice-wheat fields. It has been reported that the application of N fertilizers in soil is not totally consumed by the crop; consequently, it is more vital to enhance fertilizer usage efficiency, which can significantly reduce N_2O emissions [8]. A potential technique for reducing N_2O emissions is to reduce the amount of N input into the soil [32]. This is due to lesser N input in soil causing competition between plants and soil microorganisms, which favors soil N uptake by plants, resulting in lower N_2O emission than with high N fertilizer application. Bordoloi et al. [24] observed that reducing fertilizer rates by 25% (from 60 to 45 kg N ha^{-1}) significantly reduced N_2O emissions from fertilized rice fields. The N application method can also have an impact on N_2O production. In fact, placing N near the roots boosted NUE and lowered N_2O emissions [33]. Furthermore, optimizing N fertilizer application to better match nutrient availability with crop demand considerably reduced soil residual N, lowering N_2O emissions [34]. Split fertilizer applications at different crop stages ensure continuous N availability, which enhances NUE and decreases N_2O emissions [35].

4.2.2 Right time of fertilizer application

The right time implies applying fertilizer when the plant will benefit the most and avoiding times when fertilizer will be lost to the environment. In terms of lowering N_2O emissions, the time of fertilizer application is closely related to the amount of fertilizer used. Fertilizer application weeks after planting rather than before sowing enhances the likelihood that applied N will end up in crop tissues rather than being lost to the atmosphere and ground water.

4.2.3 Improving N fertilizer placement

Improved N placement strategies, such as urea deep placement (UDP) at a soil depth of 7 ± 10 cm, boost NUE and crop yields while lowering emissions when compared to broadcast application [36]. In flooded rice fields, UDP keeps N in the root zone as NH_4^+-N for a longer period of time, ensuring a constant supply of N to plants throughout the growing season. It has been observed that UDP boosts rice yields by 20%, NUE by 30%, and decreases N_2O emissions by 84% when compared to broadcast urea treatment [37]. Deep placement of N fertilizers in lowland rice resulted in an 80% reduced N_2O emission than traditional surface spreading [37]. This is because a substantial part of N was maintained in the soil for a longer period of time. The positioning of N closer to the plants reduces N_2O emissions significantly, as in the case of urea band application rather than broadcasting.

4.2.4 Selection of suitable fertilizers

Different fertilizers influence N_2O emissions due to varying levels of NH_4^+, NO_3^-, and organic carbon. The higher the level of N application, the greater the increase in N_2O emissions [38]. Higher quantities of N application significantly enhance the DNF, which increases N_2O emissions. Furthermore, types of fertilizers also influence NF and DNF process which ultimately affect the production of N_2O emissions. The use of anhydrous ammonia, for example, considerably enhanced N_2O emissions [39]. Grave et al. [40] investigated how different N sources affected N_2O emission in a maize-wheat rotation. They reported that, in comparison to the control plots, the application of urea and slurry increased N_2O emission by 33% and 46%, respectively. Bordoloi et al. [41] investigated the effects of various urea concentrations on N_2O emissions in a wheat cropping system and discovered that N_2O emissions rose concurrently with urea concentration, reaching a maximum of +174% with 100 kg N ha^{-1} from urea. Furthermore, Lebender et al. [42] examined the effect of the N source calcium-ammonium-nitrate (200, 400 kg ha^{-1}) on N_2O emission from the wheat crop. They observed that 400 kg N ha^{-1} consistently produced considerably more N_2O emissions than 200 kg N ha^{-1} over time. Higher N_2O emissions result with the application of calcium ammonium nitrate, particularly in moist soils with high OM [43]. In another study, Nayak et al. [44] discovered that substituting ammonium sulphate for urea enhances N_2O emissions. However, changes in N_2O emission from N fertilizers can be attributed to soil parameters including as texture, bulk density, pH, organic carbon, N, and microbial population [45]. Overall, the most important domain of intervention to reduce N_2O emissions is the selection and management of appropriate fertilizers.

4.2.5 Use of nitrification inhibitors or slow-release fertilizers

Enhanced-efficiency fertilizers including nitrification inhibitors (NIs), urease inhibitors (UIs), and control release fertilizers (CRF) have been developed to increased NUE. The use of NIs, such as dicyandiamide (DCD), in conjunction with urea or ammonium-based fertilizers (at the optimal N rate), could boost NUE while decreasing N_2O emissions in a variety of agricultural systems [46]. The NI decreases N_2O emissions directly by inhibiting NF, as well as indirectly by reducing NO_3^- availability for DNF without compromising yield [47]. The chemical components in the NI inhibit the enzymes involved for the first step of NF (ammonia mono-oxygenase; AMO), allowing NH_4^+ to remain in soils for extended periods of time [48]. As a result, the NI reduces the rates of NF and the availability of substrates for denitrifiers, lowering N_2O emissions from fertilizers [49]. Various authors observed a considerable reduction in N_2O emission with the application of various NI, including dicyandiamide, hydroquinol, nitropyrimidine, and benzoic acid [50]. Plant-derived products, such as neem oil, neem cakes, and karanja seed extract, can also be used to inhibit NF. CRF should be used in places where the sensitivity to N losses is significant [51]. CRF treatment reduced N_2O losses and N application rate in

Crops	Inorganic fertilizers management	N_2O mitigation potential	References
Wheat	Controlled-release fertilizers	29–66%	[54]
Wheat	Polymer-coated urea, sulfur-coated urea and urea-formaldehyde	39.45%, 30.74%, 11.68%	[55]
Rice	Carbon-based slow-release fertilizer	36.69%	[56]
Rice	Dicyandiamide nitrification inhibitor, Urea deep placement	95%, 73%	[57]
Rice	25% reduction in fertilizer rate (30 kg N ha^{-1}) over normal rate (40 kg N ha^{-1})	6.90–7.59%	[58]
Wheat	Urease inhibitor + urea	56.4%	[59]
Wheat	Rescheduled fertilizer N topdressings with moderate N (25 kg ha^{-1}) at sowing and remaining N dose in two equal splits	32.4%	[60]

Table 2.
Different inorganic fertilizers management and N_2O mitigation potential from rice-wheat fields.

paddy rice by 26–50% without impacting yield [52]. However, CRF can be used as a sustainable strategy to minimize N losses in conjunction or as an alternative to urea [53]. Different inorganic fertilizers management and N_2O mitigation potential from rice-wheat fields are showed in **Table 2**.

4.3 Organic fertilizer management

Organic fertilizers (OFs) such as biochar, manure, compost etc., offer soil bacteria with a variety of C compounds with diverse chemical compositions, ranging from labile to recalcitrant, that they can use to improve their growth rates and biomass during the mineralization process. OFs have dramatic, short- and long-term effects on the soil microbiome and are critical for soil health by increasing microbial activity, microbial interactions, and nutrient cycling [61]. Application and potential of different organic based fertilizers for mitigating N_2O emission from rice wheat soil have been discussed below as well as shown in **Table 3**.

4.3.1 Biochar application

Recently, the use of biochar has been regarded as an effective method for improving soil fertility, agricultural productivity, and mitigating GHG emissions from soil [19]. Biochar contains unique properties such as a highly porous structure, C-rich fine grain and enhanced surface area [70], which can draw attention to an effective GHG mitigation technique [71]. Several research have been reported by various authors relating to the amendment of biochar and its impact on GHG generation [72]. Biochar has been shown to minimize N_2O emissions by inhibiting NF and DNF processes or by promoting N_2O reduction in soil. Recent meta-analyses have revealed that biochar reduces N_2O emissions after application by an average of 20% [39]. Another study found that using biochar reduced N_2O and NH_3 emissions by 16.10% and 89.60%, respectively, as compared to a control treatment in rice crops [65]. Zhang et al. [69] reported that amendment of biochar at the rate of 10 t ha^{-1} and 40 t ha^{-1} significantly reduced the N_2O emission by 58% and 74%, respectively when compared to field without biochar application. The use of biochar raises soil pH and causes N_2O to be

Crops	Organic fertilizer management	N_2O mitigation potential	References
Wheat	Organic manure alone	39.4%	[62]
Rice wheat	Straw return + earthworm addition	19%	[63]
Wheat	Reduce N (140.3 kg ha^{-1}) + 10 t ha^{-1} biochar	7.57–12.93%	[64]
Rice-wheat	Straw biochar application	16.10%	[65]
Rice	Urea with organic amendments (poultry manure, crop residues, green manure)	11–24%	[66]
Rice	Sugarcane bagasse	31%	[67]
Rice	Rice straw + green manure	38%	[68]
Rice	Biochar at the rate of 40 t ha^{-1}	21.5%	[19]
Rice	Biochar at the rate of 10 t ha^{-1}, 40 t ha^{-1}	58, 74	[69]

Table 3.
Different organic fertilizers management and N_2O mitigation potential from rice-wheat fields.

completely converted to N_2, lowering N_2O emissions [73]. However, the effect of biochar on N_2O emissions varies depending on the amount of biochar used and soil parameters such as pH, C:N ratio, organic carbon, water status and microbial and enzymatic activity [74].

4.3.2 Use of organic amendments

Organic amendments (OA), which include compost, vermicompost, green manure, animal wastes (i.e., manures and slurries), etc., have been widely employed to reduce N fertilizer application, improve soil fertility and mitigate environmental deterioration [75]. Some studies have shown that OA increases N_2O emissions through DNF by acting as an energy source for denitrifiers and promoting the establishment of anaerobic micro-sites within soil aggregates [76]. Other researchers, on the other hand, found that OA reduces N_2O emissions by boosting N microbial absorption, reducing the availability of N substrates for N_2O synthesis via NF and DNF [77]. A long-term study found that the amount of OA is crucial for organic carbon accumulation and the consequent impact on N_2O emissions [78]. Furthermore, it is considered that the synthetic fertilizer substitution ratio by OA is a significant aspect in regulating N_2O emissions [78]. Application of fermented manures a type of OA can minimize GHG emissions due to the rapid depletion of OM pools during fermentation [79]. Nayak et al. [44] exposed that using composted manure reduced N_2O emissions considerably. In paddy soil, application of compost reduced N_2O emissions by more than 50% when compared to urea [80]. When compared to fresh straw, the use of organic material produced by aerobic composting of rice straw significantly reduced N_2O emissions [81], indicating that this strategy is environmentally favorable. Type of OA i.e., vermicomposting is a promising method that involves converting organic waste into compost in the presence of earthworms [82]. Because of the abundance of suitable resources, the material created as a result of their action has good structure and microbiological activity. In a rice study, the use of vermicompost reduced the transfer of NH_4^+ and NO_3^- to water [83]. In contrast, the combined application of biochar and vermicompost impacted soil characteristics by increasing the abundance of nosZ genes and decreasing N_2O emission [84]. As a result, combining biochar with vermicompost may be a potential way to reducing N_2O emissions. Compost or manure which is another type of OA, can help to enhance soil structure and nutrient availability to growing crops, reducing the demand for mineral fertilizer and thereby lowering GHG emissions [85]. Green manure crops such as Cowpea, Sesbania, Azzola, and Mungbean had a high ability to reduce N_2O in rice fields [86]. Because of the gradual release of nitrogen from decaying green manure residue, plant uptake efficiency and crop production can be better aligned, while N leaching losses are decreased. Different organic fertilizers management and N_2O mitigation potential from rice-wheat fields are showed in **Table 3**.

4.4 Crop management practices

4.4.1 Selection of plant cultivars

The selection of suitable crop cultivars with improved resource use efficiency appears to be an auspicious and environmentally acceptable technique for minimizing N_2O emissions from soil. Before selecting suitable crop cultivars, it is more important to investigate the mechanism of exudate and aerenchyma effects under field conditions, because variations among different types of crop cultivars have been linked to deviations in N_2O emission production, oxidation, and transport capacities [87]. According to Baruah et al. [88], different rice cultivars have varying capacities for transporting N_2O from paddy soil to the atmosphere, and these approaches are suitable for lowering GHG emissions. The physiological and anatomical properties of different rice cultivars may influence N_2O emission. Rice plant shape and physiology regulate GHG emissions by giving energy sources to microorganisms via sloughed-off root cap [89]. Another study found that lower N_2O emissions were associated with a plant strategy defined by more effectively N absorption [90]. Plant cultivars with higher N uptake were demonstrated to be able to reduce the N pool, particularly NO_3^-, resulting in lesser substrate availability for denitrifiers and, as a result, lower N_2O emission. Variation in N_2O emission among cultivars has also been documented in grain and legume intercropping [91]. In another study, researchers observed that plants contribute significantly to N_2O emissions and proposed that N_2O emission is significantly controlled by plant characteristics in the soil-crop system [92].

4.4.2 Modifying cropping schemes

In paddy field, switching from conventional puddled transplanted rice (TPR) system to directly seeded rice (DSR) may contribute to reducing GHG emissions. Under the DSR method rice seeds are sown directly in the soil where they will grow instead of transplanting seedlings. DSR methods are classified as wet (pre-germinated seeds) or dry seeding. Wet sowing method involves broadcasting pre-germinated seeds into a puddled and leveled field that is free of standing water. However, standing water on the soil surface in conventional rice fields hinders the passage of oxygen from the atmosphere into the soil and microbial activities render the water-saturated soil practically devoid of oxygen, resulting in anaerobic conditions. Denitrification is the primary mechanism for N_2O emission in TPR, because of the anaerobic conditions. In DSR, the main mechanism for N_2O emission is nitrification, which takes place under aerobic condition. In fact, it was noticed that DSR increased N_2O emission when the redox potential (RP) crossed 250 mV [93]. Therefore, in DSR water should be applied in such a way that RP be kept at a range of 100–200 mV to reduce N_2O emissions. Furthermore, it was noted that the GWP of DSR can be further reduced by converting to no-tillage farming [94]. DSR's lower GWP and higher production rate imply that it would reduce N_2O emissions. More extensive research involving GHG measurements under the concurrent effects of elements like as water, tillage, fertilizers, and biochar are, however, desperately needed to validate DSR as a feasible method that also minimizes the environmental impact.

4.5 Integrated nutrient management

Integrated nutrient management (INM) is the application of OA and inorganic fertilizers together to promote NUE and reduce N losses by coordinating crop demand with soil nutrient availability [36, 75]. Different components of INM are given in **Figure 5**. Some researchers compared the effects of NPK fertilizer, compost, and their combination on N_2O emissions [36, 95]. They exposed that combining NPK and compost lowered N_2O emissions when compared to using only compost or NPK. Furthermore, they proposed that applying composted material with a C:N ratio less than 20 considerably reduced N_2O emissions due to the release of less N during soil decomposition. The longer breakdown of C and N, as well as the slower release of mineralized N, resulted in decreased N_2O emissions when OA was used [96]. Huang et al. [97] observed a reduction in N_2O emission with increasing C:N ratio plant amendments and observed that this relationship grows stronger with the addition of inorganic N. In line with the previous findings, study found that applying OA with a lower C:N ratio alone or OA with a higher C:N ratio in combination with

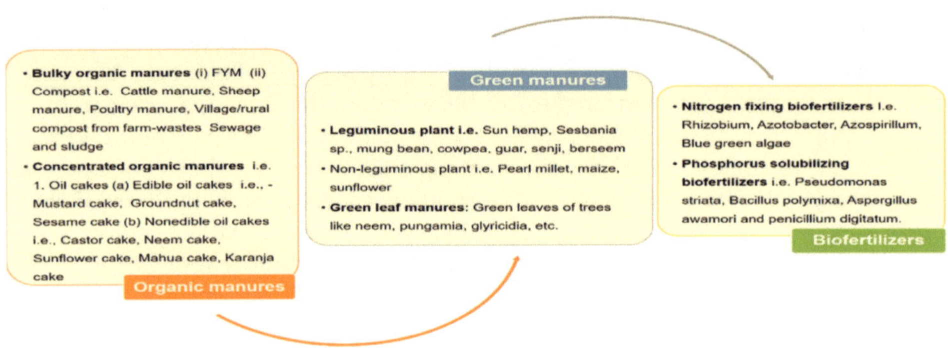

Figure 5.
Different components of INM practices.

Crops	INM practices	N_2O mitigation potential	References
Rice	Biochar (50 t ha^{-1}) + fertilizer	18%	[99]
Wheat	Chemical fertilizer reduction + organic manure	42%	[62]
Rice	Inorganic fertilizer + green manuring (mungbean)	17%	[86]
Rice	50% urea +50% poultry manure	11–14%	[66]
Rice	60 kg urea +30 kg Azolla	27.13%	[100]

Table 4.
Different INM practices and N_2O mitigation potential from rice-wheat fields.

inorganic fertilizers reduces N_2O emissions without affecting crop productivity [98]. Application and potential of INM practices for mitigating N_2O emission from rice wheat soil are shown in **Table 4**.

5. Recent technological advancements and innovations in mitigation strategies of N_2O emissions

Several technological advancements and innovations have shown promise in further reducing N_2O emissions in agriculture. While there might have been additional developments beyond that date, here are some of the notable advancements up to that point: (a) Precision agriculture technologies, such as GPS-guided equipment and sensor-based systems, enable farmers to apply fertilizers more efficiently and accurately. By precisely matching nutrient application to crop needs, these technologies can reduce nitrogen losses and subsequent N_2O emissions. (b) Efficient irrigation systems, such as drip irrigation and sensor-based watering, can optimize water and nutrient application, reducing excess nitrogen leaching and subsequent N_2O emissions. (c) Advancements in data analytics, remote sensing, and artificial intelligence can provide farmers with valuable insights into soil health, crop performance, and weather patterns. Access to real-time data can help optimize nitrogen management, leading to reduced N_2O emissions. It is essential to note that while these technological advancements hold promise in mitigating N_2O emissions, their effectiveness can vary depending on local conditions, farming practices, and the scale of implementation.

6. Adoption of greenhouse technology for climate control

The greenhouse cultivation for field crops comprises basis climate control parameters which depend on their design and complexity. It provides more or less climate control condition for plant growth and productivity [101]. This technology is beneficial in increasing crop production with limited resources and in harsh climate. Elimination of heat load is the main concern for greenhouse climate management basically in arid and semi-arid region and this can be done by reducing incoming solar radiation; removal of extra heat through air exchange; and increasing the fraction of energy partitioned into latent heat [102]. Considering shortage of resources, climate change, urbanization and population growth, the active smart greenhouse technology can support the countries food security while meeting the sustainability [103]. The current technology like fertigation, closed hydroponics, climate control systems (natural and forced ventilation, heating and fog systems and fan and pad systems) are used in greenhouses for sustainable production.

7. N_2O measurement techniques from soil

N_2O emissions from soil are largely affected by environmental variables such as substrate availability, redox potential and temperature etc., across various temporal and spatial scales. Therefore, it is necessary to understand the environmental variability of N_2O emissions, to further quantify the scale of soil–atmosphere N_2O exchange and create statistically viable measurement programmes to establish emission rates from plot to regional levels. The optimal method should be selected from the viewpoint of cost, required accuracy, time consumption, and so on. Here we describe different N_2O emission measurement techniques used by different researchers.

7.1 Closed chamber technique

The closed chamber technique is now the most extensively used measurement technique for estimating soil N_2O emissions. This is simple to use, inexpensive and allows us to study treatment effects as well as to carry out specific process studies. The closed chamber is made of 6 mm thick acrylic transparent sheets (50 cm length, 30 cm width and 70/90/120 cm height) used for gas sampling [24]. In each sampling plot, U-shaped aluminum channels (50 cm × 30 cm) is inserted into soil to a depth of 15 cm well in advance to accommodate the chambers. The chamber is placed on the U-shaped channels at the time of sampling. During gas sampling the aluminum channel is filled with water, which acted as air seal when the chamber is placed on the channel. Air inside the chamber is thoroughly mixed or homogenized with a battery-operated fan before sampling. Air temperature inside the chamber and soil temperature at 5 cm depth is measured by using mercury thermometers while taking gas samples. Gas samples are collected from the chambers by airtight syringe (50 ml volume) fitted with a three-way stop cork at an interval of 15 min (0, 15, 30 and 45 min). Gas samples are brought to the laboratory immediately after sampling and analyzed for N_2O concentrations using gas chromatograph (GC). However, there are several advantages to using a closed chamber technique, such as shortcomings related to environmental conditions (e.g., temperature effects, soil compaction, plant damage, disturbance of diffusion gradients [104], limited coverage of soil surfaces (usually less than 1 m^2), which means that spatial heterogeneity is often not adequately addressed, collar insertion in the soil and root cutting, or temporal coverage of measurements [105].

7.2 Fast-box method

The fast-box approach is a new method that will be used to investigate spatial variability of trace gas fluxes [106]. An N_2O analyzer (e.g. Tunable Diode Laser (TDL)) is coupled to a chamber in this setup. This allows for a large reduction in closure times, allowing chamber positions to be altered in minutes and spatial variability to be investigated. Closure durations of 30–60 min are usual with standard GC procedures.

7.3 Micrometeorological measurements

Micrometeorological measurement of N_2O with TDL detection is based on the principle of diode laser absorption spectroscopy. It offers a non-intrusive, continuous spatially integrated measurement technique for detecting and quantifying baseline and episodic N_2O emissions at the paddock scale. Pattey et al. [107], analyzed the wide variety of conceivable micrometeorolgical applications of TDL technology. The TDL measurements were made using the TGA-100A (Campbell Scientific Inc.). They were reported that dried air was sampled from the two heights at 3 s intervals, raw N_2O measurements were taken at 10 Hz, and concentration data were averaged over 20 min. Micrometeorological approaches require homogeneous areas with a considerable fetch (>1 hectare) that are unaffected by structures, trees, hills, and other factors. For the straight fetch area, land use, land management, vegetation, and soil qualities should be uniform. These methods are most commonly used in flat terrain with vast, homogeneous land uses, such as pasture, grassland, maize, or wheat monocrops, woods, or tree plantations.

7.4 Modeling based approaches

Over the last few decades, a wide variety of process models for modeling soil N_2O emissions have been created, each of which is suitable to one or more specific ecosystem types (e.g., arable, grassland, forest) [108]. Models can be classified depending on their degree of complexity of the biogeochemical N cycle such as mineralization, nitrification, denitrification as well as trace gas production, consumption and emission processes.

8. Role of policies and economic incentives in promoting N_2O mitigation strategies

Policy formulation should aim to encourage farmers to adopt mitigation methods that do not compromise their productivity and profitability. To promote the use of mitigation technology in agriculture, three main paths should be pursued: investments, incentives, and information. Agricultural output as a GHG source is unique due to its small-scale, dispersed nature, and often inadequate physical and institutional infrastructure. Policy initiatives should consider these variations and implement cost-effective payment schemes to incentivize and support agricultural mitigation efforts. Establish an extension system to assist farmers in adopting climate change mitigation practices. This support can include facilitating access to new markets, especially carbon markets, providing information on new regulatory systems, and informing farmers about government goals and policies related to climate change. Increase research funding to enhance our understanding of how climate change impacts agriculture. This includes studying the interactions between climate change and agricultural practices, which can lead to better forecasts and informed policies for long-term sustainable growth, particularly with a focus on pro-poor development. By implementing these policy approaches, the government can effectively encourage farmers to adopt mitigation methods that contribute to climate change mitigation while ensuring their agricultural productivity and economic well-being.

9. Co-benefits and potential trade-offs associated with N_2O mitigation strategies

N_2O mitigation strategies in agriculture can offer both co-benefits and potential trade-offs. These strategies aim to reduce nitrous oxide emissions from agricultural practices, thereby addressing its negative impact on climate change and the environment. However, the effectiveness of these strategies may vary, and they can have additional implications for agricultural productivity, soil health, and economic aspects. There are several co-benefits associated with N_2O mitigation strategies. By implementing N_2O mitigation strategies, such as better nitrogen management practices, farmers can help reduce greenhouse gas emissions and contribute to global efforts to combat climate change. Some N_2O mitigation strategies, such as using cover crops, reduced tillage, and organic farming practices, can enhance soil health. These practices can increase soil organic matter, improve nutrient cycling, and enhance soil structure, leading to better water retention and reduced erosion. Implementing N_2O mitigation measures often involves optimizing nitrogen use on farms. This can lead to better nitrogen use efficiency, which benefits farmers economically by reducing input costs and minimizing nitrogen losses to the environment. N_2O is not the only nitrogen compound emitted from agricultural practices. Nitrogen runoff and leaching can lead to water pollution, affecting aquatic ecosystems and human water supplies. N_2O mitigation strategies can also reduce other forms of nitrogen pollution, thereby improving water quality. Beside these co-benefits there are also some potential trade-offs associated with N_2O mitigation strategies. Some N_2O mitigation strategies, particularly those that involve reducing synthetic nitrogen fertilizers, can lead to decreased crop yields if not managed properly. Balancing nitrogen inputs to optimize both yield and environmental benefits can be challenging. Implementing certain N_2O mitigation strategies may involve initial investments in new technologies or changes in farm management practices, which can impose additional costs on farmers. While some practices may have long-term economic benefits, short-term financial constraints can be a trade-off. Agricultural systems are complex, and the effectiveness of N_2O mitigation strategies can vary depending on factors such as soil type, climate, and local management practices. The uncertainty associated with their outcomes can be a trade-off.

10. Knowledge and capacity building in promoting N_2O mitigation strategies

The adoption of N_2O mitigation strategies in agriculture requires more than just the availability of technologies and practices. Awareness campaigns, training programs, and knowledge-sharing platforms play a critical role in promoting the understanding and adoption of these strategies among farmers and stakeholders. These initiatives can address barriers to adoption, disseminate valuable information, and foster behavioral change toward sustainable agricultural practices. Many farmers might not be aware of the environmental impact of N_2O emissions or the available mitigation strategies. Awareness campaigns can help disseminate knowledge about the link between agricultural practices, GHGs emissions, and climate change, thus creating a sense of urgency and responsibility among farmers. Training programs provide farmers and agricultural stakeholders with the necessary skills and knowledge to implement N_2O mitigation strategies effectively. These programs can cover various topics, such as precision agriculture, improved fertilizer management, and soil health practices. Farmers might be hesitant to adopt new technologies due to unfamiliarity or uncertainty about their benefits. Knowledge-sharing platforms can showcase successful case studies, demonstrations, and testimonials from other farmers who have successfully implemented N_2O mitigation practices. Different regions and farming systems have varying challenges and opportunities for N_2O mitigation. Awareness campaigns and knowledge-sharing platforms can tailor information and strategies to suit specific contexts, making it more relevant and applicable for farmers. Overall, fostering awareness, providing relevant training, and establishing knowledge-sharing platforms are essential components of promoting the adoption of N_2O mitigation strategies in agriculture.

11. Conclusions

It is becoming obvious that no single management strategies can result in increased crop yields and lower N_2O emissions across the wide geographical areas. While site-to-site variability and climate influences on N_2O emissions are significant, site-specific adjustments in agricultural management strategies can provide remedies and should be given more attention. Understanding the mechanisms of N_2O formation in rice-wheat fields has led to the development of various mitigation techniques to reduce N_2O emissions. Site-specific fertilizer management, modifying irrigation strategies such as AMD, intermittent irrigation and the use of DSR all help to reduce N_2O emissions. N_2O emissions can be reduced by using fermented manures, altering N fertilizer sources, timing, placement methods, applying NI, or using slow-release fertilizers. Similarly, biochar, compost, straw ash inclusion, and INM have the ability to significantly reduce N_2O emissions while maintaining crop production. On the other hand, farmers will only accept mitigation techniques that do not reduce grain yield. More agricultural focus may be drawn to site-specific management adjustments and the use of technologies that will assist limit N losses via ammonia volatilization and nitrate runoff, leaching, and drainage pathways. The mitigation measures outlined above are scientific discoveries, but effective implementation of these options alone or in combination at the farmer level requires a deliberate policy and strong government backing. The policy to reduce or eliminate N_2O emissions into the atmosphere will differ depending on the region or country and it will be heavily reliant on government financial assistance. However, in order for such techniques to be effective and fruitful in reducing GHG emissions and maintaining crop output in a changing environment, all social, economic, educational, and political barriers must be addressed. More research on climate-smart agriculture is needed to validate at the agricultural system level and to inform policymakers about the projected implications of climate change and the effectiveness of mitigation strategies.

Acknowledgements

Authors are grateful to SERB (File No. EEQ/2018/000125), GOI, New Delhi, India, for financial support and Central University of Jharkhand, Brambe, Ranchi, India, for facilitating and supporting the activities.

Conflict of interest

The authors declare no conflict of interest.

Author details

Raushan Kumar and Nirmali Bordoloi*
Department of Environmental Sciences, Central University of Jharkhand, Brambe, Ranchi, India

*Address all correspondence to: nirmali.bordoloi@cuj.ac.in

References

[1] Lynch J, Cain M, Frame D, Pierrehumbert R. Agriculture's contribution to climate change and role in mitigation is distinct from predominantly fossil CO_2-emitting sectors. Frontiers in Sustainable Food Systems. 2021;**4**:518039. DOI: 10.3389/fsufs.2020.518039

[2] Malyan SK, Bhatia A, Fagodiya RK, Kumar SS, Kumar A, Gupta DK, et al. Plummeting global warming potential by chemicals interventions in irrigated rice: A lab to field assessment. Agriculture, Ecosystems & Environment. 2021;**319**:107545

[3] Ravishankara AR, Daniel JS, Portmann RW. Nitrous oxide (N_2O): The dominant ozone-depleting substance emitted in the 21st century. Science. 2009;**326**:123-125. DOI: 10.1126/science.1176985

[4] Change IC. Mitigation of climate change. Contribution of Working Group III to the Fifth Assessment Report of the Intergovernmental Panel on Climate Change. 2014;**1454**:147

[5] Vries DW. Impacts of nitrogen emissions on ecosystems and human health: A mini review. Current Opinion in Environmental Science & Health. 2021;**21**:100249. DOI: 10.1016/j.coesh.2021.100249

[6] Delmotte MV, Zhai P, Pirani A, Connors SL, Péan C, Berger S, et al. Climate change 2021: The physical science basis. Contribution of Working Group I to the Sixth Assessment Report of the Intergovernmental Panel on Climate Change;**2021**:2

[7] Tian H, Xu R, Canadell JG, Thompson RL, Winiwarter W, Suntharalingam P, et al. A comprehensive quantification of global nitrous oxide sources and sinks. Nature. 2020;**586**:248-256

[8] Gupta K, Kumar R, Baruah KK, Hazarika S, Karmakar S, Bordoloi N. Greenhouse gas emission from rice fields: A review from Indian context. Environmental Science and Pollution Research. 2021;**28**:30551-30572. DOI: 10.1007/s11356-021-13935-1

[9] Mbow C, Rosenzweig C, Barioni LG, Benton TG, Herrero M, Krishnapillai M, et al. Food security. In: Shukla PR, Skea J, Buendia EC, Masson-Delmotte V, Pörtner H-O, Roberts DC, et al., editors. Climate Change and Land: An IPCC Special Report on Climate Change, Desertification, Land Degradation, Sustainable Land Management, Food Security, and Greenhouse Gas Fluxes in Terrestrial Ecosystems. Intergovernmental Panel on Climate Change; 2019. pp. 1-18

[10] Kammann C, Ippolito J, Hagemann N, Borchard N, Cayuela ML, Estavillo JM, et al. Biochar as a tool to reduce the agricultural greenhouse-gas burden – Knowns, unknowns and future research needs. Journal of Environmental Engineering and Landscape Management. 2017;**25**:114-139. DOI: 10.3846/16486897.2017.1319375

[11] Tian L, Zhu B, Akiyama H. Seasonal variations in indirect N_2O emissions from an agricultural headwater ditch. Biology and Fertility of Soils. 2017;**53**:651-662. DOI: 10.1007/s00374-017-1207-z

[12] Pilegaard K. Processes regulating nitric oxide emissions from soils. Philosophical Transactions of the Royal Society B: Biological Sciences.

2013;**368**:20130126. DOI: 10.1098/rstb.2013.0126

[13] Braker G, Conrad R. Diversity, structure, and size of N_2O-producing microbial communities in soils-what matters for their functioning? In: Laskin AI, Sariaslani S, Gadd GM, editors. Advances in Applied Microbiology. Vol. 75. Amsterdam, The Netherlands: Academic Press; 2011. pp. 33-70. DOI: 10.1016/B978-0-12-387046-9.00002-5

[14] Wang M, Fu Y, Wang Y, Li Y, Shen J, Liu X, et al. Pathways and mechanisms by which biochar application reduces nitrogen and phosphorus runoff losses from a rice agroecosystem. Science of the Total Environment. 2021;**797**:149193. DOI: 10.1016/j.scitotenv.2021.149193

[15] Baggs EM. Soil microbial sources of nitrous oxide: Recent advances in knowledge, emerging challenges and future direction. Current Opinion in Environmental Sustainability. 2011;**3**:321-327. DOI: 10.1016/j.cosust.2011.08.011

[16] Liu S, Zhang Y, Lin F, Zhang L, Zou J. Methane and nitrous oxide emissions from direct-seeded and seedling-transplanted rice paddies in Southeast China. Plant and Soil. 2014;**374**:285-297. DOI: 10.1007/s11104-013-1878-7

[17] Sanz-Cobena A, Lassaletta L, Aguilera E, Prado A d, Garnier J, Billen G, et al. Strategies for greenhouse gas emissions mitigation in Mediterranean agriculture: A review. Agriculture, Ecosystems & Environment. 2017;**238**:5-24. DOI: 10.1016/j.agee.2016.09.038

[18] Trost B, Prochnow A, Drastig K, Meyer-Aurich A, Ellmer F, Baumecker M. Irrigation, soil organic carbon and N2O emissions. A review. Agronomy for Sustainable Development. 2013;**33**:733-749. DOI: 10.1007/s13593-013-0134-0

[19] Shihong Y, Xiao Y, Sun X, Ding J, Jiang Z, Xu J. Biochar improved rice yield and mitigated CH_4 and N_2O emissions from paddy field under controlled irrigation in the Taihu Lake Region of China. Atmospheric Environment. 2019;**200**:69-77. DOI: 10.1016/j.atmosenv.2018.12.003

[20] Rassaei F. Nitrous oxide emissions from rice paddy: Impacts of rice straw and water management. Environmental Progress & Sustainable Energy. 2023;**n/a**:e14066. DOI: 10.1002/ep.14066

[21] Zhang H, Liang Q, Peng Z, Zhao Y, Tan Y, Zhang X, et al. Response of greenhouse gases emissions and yields to irrigation and straw practices in wheat-maize cropping system. Agricultural Water Management. 2023;**282**:108281. DOI: 10.1016/j.agwat.2023.108281

[22] Kar S, Pramanick B, Brahmachari K, Saha G, Mahapatra BS, Saha A, et al. Exploring the best tillage option in rice based diversified cropping systems in alluvial soil of eastern India. Soil and Tillage Research. 2021;**205**:104761. DOI: 10.1016/j.still.2020.104761

[23] Chaudhuary P, Chudhury SR, Das A, Mandal J, Ghosh M, Acharya S, et al. Productivity, profitability and greenhouse gas Emission from Rice-wheat cropping system under different tillage and nitrogen management practices. Indian. Journal of Agricultural Research. 2020;**54**:285-292. DOI: 10.18805/IJARe.A-5325

[24] Bordoloi N, Baruah KK, Bhattacharyya P, Gupta PK. Impact of nitrogen fertilization and tillage practices on nitrous oxide emission from a summer rice ecosystem. Archives of Agronomy

and Soil Science. 2019;**65**:1493-1506. DOI: 10.1080/03650340.2019.1566716

[25] Gupta DK, Bhatia A, Kumar A, Das TK, Jain N, Tomer R, et al. Mitigation of greenhouse gas emission from rice–wheat system of the Indo-Gangetic plains: Through tillage, irrigation and fertilizer management. Agriculture, Ecosystems & Environment. 2016;**230**:1-9. DOI: 10.1016/j.agee.2016.05.023

[26] Oorts K, Merckx R, Gréhan E, Labreuche J, Nicolardot B. Determinants of annual fluxes of CO_2 and N_2O in long-term no-tillage and conventional tillage systems in northern France. Soil and Tillage Research. 2007;**95**:133-148. DOI: 10.1016/j.still.2006.12.002

[27] Six J, Conant RT, Paul EA, Paustian K. Stabilization mechanisms of soil organic matter: Implications for C-saturation of soils. Plant and Soil. 2002;**241**:155-176. DOI: 10.1023/A:1016125726789

[28] van Kessel C, Venterea R, Six J, Adviento-Borbe MA, Linquist B, van Groenigen KJ. Climate, duration, and N placement determine N_2O emissions in reduced tillage systems: A meta-analysis. Global Change Biology. 2013;**19**:33-44. DOI: 10.1111/j.1365-2486.2012.02779.x

[29] Gregorutti VC, Caviglia OP. Nitrous oxide emission after the addition of organic residues on soil surface. Agriculture, Ecosystems & Environment. 2017;**246**:234-242. DOI: 10.1016/j.agee.2017.06.016

[30] Sander BO, Samson M, Buresh RJ. Methane and nitrous oxide emissions from flooded rice fields as affected by water and straw management between rice crops. Geoderma. 2014;**235-236**:355-362. DOI: 10.1016/j.geoderma.2014.07.020

[31] Burger M, Jackson LE. Microbial immobilization of ammonium and nitrate in relation to ammonification and nitrification rates in organic and conventional cropping systems. Soil Biology and Biochemistry. 2003;**35**:29-36. DOI: 10.1016/S0038-0717(02)00233-X

[32] Saha G, Kar B, Karmakar S. Methane and nitrous oxide emission from kharif rice field as influenced by nutrients and moisture regimes in new alluvial agroclimatic region of West Bengal. Current Science. 2017;**112**:989-995

[33] Haider A, Bashir A, ul Husnain MI. Impact of agricultural land use and economic growth on nitrous oxide emissions: Evidence from developed and developing countries. Science of the Total Environment. 2020;**741**:140421. DOI: 10.1016/j.scitotenv.2020.140421

[34] Sharma LK, Bali SK. A review of methods to improve nitrogen use efficiency in agriculture. Sustainability (Switzerland). 2018;**10**(1):51. DOI: 10.3390/su10010051

[35] An H, Owens J, Beres B, Li Y, Hao X. Nitrous oxide emissions with enhanced efficiency and conventional urea fertilizers in winter wheat. Nutrient Cycling in Agroecosystems. 2021;**119**:307-322. DOI: 10.1007/s10705-021-10118-9

[36] Yao Z, Zheng X, Wang R, Liu C, Lin S, Butterbach-Bahl K. Benefits of integrated nutrient management on N2O and NO mitigations in water-saving ground cover rice production systems. Science of the Total Environment. 2019;**646**:1155-1163. DOI: 10.1016/j.scitotenv.2018.07.393

[37] Gaihre YK, Singh U, Islam SMM, Huda A, Islam MR, Satter MA, et al. Impacts of urea deep placement on nitrous oxide and nitric oxide emissions

from rice fields in Bangladesh. Geoderma. 2015;**259-260**:370-379. DOI: 10.1016/j.geoderma.2015.06.001

[38] Shakoor A, Xu Y, Wang Q, Chen N, He F, Zuo H, et al. Effects of fertilizer application schemes and soil environmental factors on nitrous oxide emission fluxes in a rice-wheat cropping system, East China. PLoS One. 2018;**13**:1-16. DOI: 10.1371/journal.pone.0202016

[39] Shakoor A, Shakoor S, Rehman A, Ashraf F, Abdullah M, Shahzad SM, et al. Effect of animal manure, crop type, climate zone, and soil attributes on greenhouse gas emissions from agricultural soils—A global meta-analysis. Journal of Cleaner Production. 2021;**278**:124019. DOI: 10.1016/j.jclepro.2020.124019

[40] Grave RA, Nicoloso R d S, Cassol PC, da Silva MLB, Mezzari MP, Aita C, et al. Determining the effects of tillage and nitrogen sources on soil N_2O emission. Soil and Tillage Research. 2018;**175**:1-12. DOI: 10.1016/j.still.2017.08.011

[41] Bordoloi N, Baruah KK, Bhattacharyya P. Emission estimation of nitrous oxide (N_2O) from a wheat cropping system under varying tillage practices and different levels of nitrogen fertiliser. Soil Research. 2016;**54**:767-776. DOI: 10.1071/SR15268

[42] Lebender U, Senbayram M, Lammel J, Kuhlmann H. Impact of mineral N fertilizer application rates on N2O emissions from arable soils under winter wheat. Nutrient Cycling in Agroecosystems. 2014;**100**:111-120. DOI: 10.1007/s10705-014-9630-0

[43] Watson CJ, Laughlin RJ, McGeough KL. Modification of nitrogen fertilisers using inhibitors: Opportunities and potentials for improving nitrogen use efficacy. In: Proceedings 658. Vol. 658. London, United Kingdom: International Fertilizer Society; 2009. pp. 1-40

[44] Nayak D, Saetnan E, Cheng K, Wang W, Koslowski F, Cheng YF, et al. Management opportunities to mitigate greenhouse gas emissions from Chinese agriculture. Agriculture, Ecosystems and Environment. 2015;**209**:108-124. DOI: 10.1016/j.agee.2015.04.035

[45] Kelliher FM, Cox N, Van Der Weerden TJ, De Klein CAM, Luo J, Cameron KC, et al. Statistical analysis of nitrous oxide emission factors from pastoral agriculture field trials conducted in New Zealand. Environmental Pollution. 2014;**186**:63-66. DOI: 10.1016/j.envpol.2013.11.025

[46] Recio J, Vallejo A, Le-Noë J, Garnier J, García-Marco S, Álvarez JM, et al. The effect of nitrification inhibitors on NH_3 and N_2O emissions in highly N fertilized irrigated Mediterranean cropping systems. Science of the Total Environment. 2018;**636**:427-436. DOI: 10.1016/j.scitotenv.2018.04.294

[47] Cayuela ML, Aguilera E, Sanz-Cobena A, Adams DC, Abalos D, Barton L, et al. Direct nitrous oxide emissions in Mediterranean climate cropping systems: Emission factors based on a meta-analysis of available measurement data. Agriculture, Ecosystems and Environment. 2017;**238**:25-35. DOI: 10.1016/j.agee.2016.10.006

[48] Gilsanz C, Báez D, Misselbrook TH, Dhanoa MS, Cárdenas LM. Development of emission factors and efficiency of two nitrification inhibitors, DCD and DMPP. Agriculture, Ecosystems & Environment. 2016;**216**:1-8. DOI: 10.1016/j.agee.2015.09.030

[49] Zebarth BJ, Snowdon E, Burton DL, Goyer C, Dowbenko R. Controlled

release fertilizer product effects on potato crop response and nitrous oxide emissions under rain-fed production on a medium-textured soil. Canadian Journal of Soil Science. 2012;**92**:759-769. DOI: 10.4141/CJSS2012-008

[50] Akiyama H, Yan X, Yagi K. Estimations of emission factors for fertilizer-induced direct N2O emissions from agricultural soils in Japan: Summary of available data. Soil Science and Plant Nutrition. 2006;**52**:774-787. DOI: 10.1111/j.1747-0765.2006.00097.x

[51] Nawaz M, Chattha MU, Chattha MB, Ahmad R, Munir H, Usman M, et al. Assessment of compost as nutrient supplement for spring planted sugarcane (*Saccharum officinarum* L.). Journal of Animal and Plant Sciences. 2017;**27**:283-293

[52] Nawaz M, Khan S, Ali H, Ijaz M, Chattha MU, Hassan MU, et al. Assessment of environment-friendly usage of spent wash and its nutritional potential for sugarcane production. Communications in Soil Science and Plant Analysis. 2019;**50**:1239-1249. DOI: 10.1080/00103624.2019.1614601

[53] Chen H, Zhao Y, Feng H, Liu J, Si B, Zhang A, et al. Effects of straw and plastic film mulching on greenhouse gas emissions in Loess Plateau, China: A field study of 2 consecutive wheat-maize rotation cycles. Science of the Total Environment. 2017;**579**:814-824. DOI: 10.1016/j.scitotenv.2016.11.022

[54] Liu Q, Liu Y, Hao X, Song C, Zong Y, Zhang D, et al. Effects of controlled-release fertilizer on N_2O emissions in wheat under elevated CO_2 concentration and temperature. Plant and Soil. 2023;**488**:343-361. DOI: 10.1007/s11104-023-05972-1

[55] Ma Q, Qian Y, Yu Q, Cao Y, Tao R, Zhu M, et al. Controlled-release nitrogen fertilizer application mitigated N losses and modified microbial community while improving wheat yield and N use efficiency. Agriculture, Ecosystems & Environment. 2023;**349**:108445. DOI: 10.1016/j.agee.2023.108445

[56] Wang X, Wang M, Chen L, Shutes B, Yan B, Zhang F, et al. Nitrogen migration and transformation in a saline-alkali paddy ecosystem with application of different nitrogen fertilizers. Environmental Science and Pollution Research. 2023;**30**:51665-51678. DOI: 10.1007/s11356-023-25984-9

[57] Gaihre YK, Singh U, Bible WD, Fugice J, Sanabria J. Mitigating N_2O and NO emissions from direct-seeded Rice with nitrification inhibitor and urea deep placement. Rice Science. 2020;**27**:434-444. DOI: 10.1016/j.rsci.2020.03.005

[58] Baruah KK, Kumar R, Bordoloi N. Identification of optimal level of nitrogen fertilizer and tillage practice to reduce nitrous oxide emissions and maximize the nitrogen use efficiency and productivity from tropical rice paddy. Paddy and Water Environment. 2022;**20**:483-497. DOI: 10.1007/s10333-022-00906-6

[59] Cheng Y, Elrys AS, Wang J, Xu C, Ni K, Zhang J, et al. Application of enhanced-efficiency nitrogen fertilizers reduces mineral nitrogen usage and emissions of both N_2O and NH_3 while sustaining yields in a wheat-rice rotation system. Agriculture, Ecosystems & Environment. 2022;**324**:107720. DOI: 10.1016/j.agee.2021.107720

[60] Singh V, Kaur S, Singh J, Kaur A, Gupta RK. Rescheduling fertilizer nitrogen topdressing timings for improving productivity and mitigating N_2O emissions in timely and late sown irrigated wheat (*Triticum aestivum* L.). Archives of Agronomy

and Soil Science. 2021;**67**:647-659. DOI: 10.1080/03650340.2020.1742327

[61] Ling N, Zhu C, Xue C, Chen H, Duan Y, Peng C, et al. Insight into how organic amendments can shape the soil microbiome in long-term field experiments as revealed by network analysis. Soil Biology and Biochemistry. 2016;**99**:137-149. DOI: 10.1016/j.soilbio.2016.05.005

[62] Gao H, Xi Y, Wu X, Pei X, Liang G, Bai J, et al. Partial substitution of manure reduces nitrous oxide emission with maintained yield in a winter wheat crop. Journal of Environmental Management. 2023;**326**:116794. DOI: 10.1016/j.jenvman.2022.116794

[63] Kan Z-R, Zhou J, Li F-M, Sheteiwy MS, Qi J, Chen C, et al. Straw incorporation interacting with earthworms mitigates N_2O emissions from upland soil in a rice-wheat rotation system. Science of the Total Environment. 2023;**859**:160338. DOI: 10.1016/j.scitotenv.2022.160338

[64] Li W, Xie H, Ren Z, Li T, Wen X, Han J, et al. Response of N_2O emissions to N fertilizer reduction combined with biochar application in a rain-fed winter wheat ecosystem. Agriculture, Ecosystems & Environment. 2022;**333**:107968. DOI: 10.1016/j.agee.2022.107968

[65] He L, Xu Y, Li J, Zhang Y, Liu Y, Lyu H, et al. Biochar mitigated more N-related global warming potential in rice season than that in wheat season: An investigation from ten-year biochar-amended rice-wheat cropping system of China. Science of the Total Environment. 2022;**821**:153344. DOI: 10.1016/j.scitotenv.2022.153344

[66] Mohanty S, Nayak AK, Swain CK, Dhal BR, Kumar A, Kumar U, et al. Impact of integrated nutrient management options on GHG emission, N loss and N use efficiency of low land rice. Soil and Tillage Research. 2020;**200**:104616. DOI: 10.1016/j.still.2020.104616

[67] Baruah A, Baruah KK. Organic manures and crop residues as fertilizer substitutes: Impact on nitrous oxide emission, plant growth and grain yield in pre-monsoon rice cropping system. Journal of Environmental Protection. 2015;**06**:755-770. DOI: 10.4236/jep.2015.67069

[68] Bhattacharyya P, Roy KS, Neogi S, Adhya TK, Rao KS, Manna MC. Effects of rice straw and nitrogen fertilization on greenhouse gas emissions and carbon storage in tropical flooded soil planted with rice. Soil and Tillage Research. 2012;**124**:119-130. DOI: 10.1016/j.still.2012.05.015

[69] Zhang A, Cui L, Pan G, Li L, Hussain Q, Zhang X, et al. Effect of biochar amendment on yield and methane and nitrous oxide emissions from a rice paddy from Tai Lake plain, China. Agriculture, Ecosystems & Environment. 2010;**139**:469-475. DOI: 10.1016/j.agee.2010.09.003

[70] Kumari K, Kumar R, Bordoloi N, Minkina T, Keswani C, Bauddh K. Unravelling the recent developments in the production technology and efficient applications of biochar for agro-ecosystems. Agriculture. 2023;**13**:512. DOI: 10.3390/agriculture13030512

[71] Zhang H-L, Bai X-L, Xue J-F, Chen Z-D, Tang H-M, Chen F. Emissions of CH_4 and N_2O under different tillage systems from double-cropped paddy fields in southern China. PLoS One. 2013;**8**:e65277

[72] Liu WJ, Jiang H, Yu HQ. Emerging applications of biochar-based materials

for energy storage and conversion. Energy and Environmental Science. 2019;**12**:1751-1779. DOI: 10.1039/c9ee00206e

[73] Cornelissen G, Rutherford DW, Arp HPH, Dörsch P, Kelly CN, Rostad CE. Sorption of pure N2O to biochars and other organic and inorganic materials under anhydrous conditions. Environmental Science & Technology. 2013;**47**:7704-7712. DOI: 10.1021/es400676q

[74] Bakken LR, Bergaust L, Liu B, Frostegård Å. Regulation of denitrification at the cellular level: A clue to the understanding of N2O emissions from soils. Philosophical Transactions of the Royal Society B: Biological Sciences. 2012;**367**:1226-1234. DOI: 10.1098/rstb.2011.0321

[75] Seleiman MF, Abdelaal MS. Effect of organic, Inorganic and bio-fertilization on growth, yield and quality traits of some chickpea (*Cicer arietinum* L.) varieties. Egyptian Journal of Agronomy. 2018;**40**:105-117. DOI: 10.21608/agro.2018.2869.1093

[76] Jain N, Arora P, Tomer R, Mishra SV, Bhatia A, Pathak H, et al. Greenhouse gases emission from soils under major crops in Northwest India. Science of the Total Environment. 2016;**542**:551-561. DOI: 10.1016/j.scitotenv.2015.10.073

[77] Gu J, Yuan M, Liu J, Hao Y, Zhou Y, Qu D, et al. Trade-off between soil organic carbon sequestration and nitrous oxide emissions from winter wheat-summer maize rotations: Implications of a 25-year fertilization experiment in Northwestern China. Science of the Total Environment. 2017;**595**:371-379

[78] Chattha MU, Hassan MU, Barbanti L, Chattha MB, Khan I, Usman M, et al. Composted sugarcane by-product press mud cake supports wheat growth and improves soil properties. International Journal of Plant Production. 2019;**13**:241-249. DOI: 10.1007/s42106-019-00051-x

[79] Doan TT, Bouvier C, Bettarel Y, Bouvier T, Henry-des-Tureaux T, Janeau JL, et al. Influence of buffalo manure, compost, vermicompost and biochar amendments on bacterial and viral communities in soil and adjacent aquatic systems. Applied Soil Ecology. 2014;**73**:78-86. DOI: 10.1016/j.apsoil.2013.08.016

[80] Zhang JB, Zhu TB, Cai ZC, Qin SW, Müller C. Effects of long-term repeated mineral and organic fertilizer applications on soil nitrogen transformations. European Journal of Soil Science. 2012;**63**:75-85. DOI: 10.1111/j.1365-2389.2011.01410.x

[81] Doan TT, Henry-des-Tureaux T, Rumpel C, Janeau J-L, Jouquet P. Impact of compost, vermicompost and biochar on soil fertility, maize yield and soil erosion in northern Vietnam: A three year mesocosm experiment. Science of the Total Environment. 2015;**514**:147-154. DOI: 10.1016/j.scitotenv.2015.02.005

[82] Aamer M, Shaaban M, Hassan MU, Guoqin H, Ying L, Hai Ying T, et al. Biochar mitigates the N_2O emissions from acidic soil by increasing the nosZ and nirK gene abundance and soil pH. Journal of Environmental Management. 2020;**255**:109891. DOI: 10.1016/j.jenvman.2019.109891

[83] Barthod J, Rumpel C, Dignac M-F. Composting with additives to improve organic amendments. A review. Agronomy for Sustainable Development. 2018;**38**:17. DOI: 10.1007/s13593-018-0491-9

[84] Kool DM, Van Groenigen JW, Wrage N. Chapter six—Source determination of nitrous oxide based on nitrogen and oxygen isotope tracing: Dealing with oxygen exchange. In: Klotz MG, Stein LY, editors. Methods in Enzymology. Vol. 496. Amsterdam, The Netherlands: Academic Press; 2011. pp. 139-160. DOI: 10.1016/B978-0-12-386489-5.00006-3

[85] Supparattanapan S, Saenjan P, Quantin C, Maeght JL, Grünberger O. Salinity and organic amendment effects on methane emission from a rain-fed saline paddy field. Soil Science and Plant Nutrition. 2009;**55**:142-149. DOI: 10.1111/j.1747-0765.2008.00330.x

[86] Bharati V, Kumar V, Mohanty S, Choudhury SR. Performance of rice-wheat cropping system under resource conservation and various nutrient management practices. International Journal of Chemistry Studies. 2020;**8**:2022-2026. DOI: 10.22271/chemi.2020.v8.i3ab.9507

[87] Lou Y, Inubushi K, Mizuno T, Hasegawa T, Lin Y, Sakai H, et al. CH_4 emission with differences in atmospheric CO_2 enrichment and rice cultivars in a Japanese paddy soil. Global Change Biology. 2008;**14**:2678-2687. DOI: 10.1111/j.1365-2486.2008.01665.x

[88] Baruah KK, Gogoi B, Gogoi P. Plant physiological and soil characteristics associated with methane and nitrous oxide emission from rice paddy. Physiology and Molecular Biology of Plants. 2010;**16**:79-91. DOI: 10.1007/s12298-010-0010-1

[89] Aulakh MS, Wassmann R, Bueno C, Rennenberg H. Impact of root exudates of different cultivars and plant development stages of rice (*Oryza sativa* L.) on methane production in a paddy soil. Plant and Soil. 2001;**230**:77-86. DOI: 10.1023/A:1004817212321

[90] Hou AX, Chen GX, Wang ZP, Van Cleemput O, Patrick WH. Methane and nitrous oxide emissions from a rice field in relation to soil redox and microbiological processes. Soil Science Society of American Journal. 2000;**64**:2180-2186. DOI: 10.2136/sssaj2000.6462180x

[91] Pathak H. Mitigating greenhouse gas and nitrogen loss with improved fertilizer management in rice: Quantification and economic assessment. Nutrient Cycling in Agroecosystems. 2010;**87**:443-454. DOI: 10.1007/s10705-010-9350-z

[92] Wegner BR, Chalise KS, Singh S, Lai L, Abagandura GO, Kumar S, et al. Response of soil surface greenhouse gas fluxes to crop residue removal and cover crops under a corn–soybean rotation. Journal of Environmental Quality. 2018;**47**:1146-1154. DOI: 10.2134/jeq2018.03.0093

[93] Lehman RM, Osborne SL, Duke SE. Diversified no-till crop rotation reduces nitrous oxide emissions, increases soybean yields, and promotes soil carbon accrual. Soil Science Society of America Journal. 2017;**81**:76-83. DOI: 10.2136/sssaj2016.01.0021

[94] Ahmad S, Li C, Dai G, Zhan M, Wang J, Pan S, et al. Greenhouse gas emission from direct seeding paddy field under different rice tillage systems in Central China. Soil and Tillage Research. 2009;**106**:54-61. DOI: 10.1016/j.still.2009.09.005

[95] Dittert K, Lampe C, Gasche R, Butterbach-Bahl K, Wachendorf M, Papen H, et al. Short-term effects of single or combined application of mineral N fertilizer and cattle slurry on the fluxes of radiatively active trace

gases from grassland soil. Soil Biology and Biochemistry. 2005;**37**:1665-1674. DOI: 10.1016/j.soilbio.2005.01.029

[96] Meng L, Ding W, Cai Z. Long-term application of organic manure and nitrogen fertilizer on N_2O emissions, soil quality and crop production in a sandy loam soil. Soil Biology and Biochemistry. 2005;**37**:2037-2045. DOI: 10.1016/j.soilbio.2005.03.007

[97] Huang Y, Zou J, Zheng X, Wang Y, Xu X. Nitrous oxide emissions as influenced by amendment of plant residues with different C:N ratios. Soil Biology and Biochemistry. 2004;**36**:973-981. DOI: 10.1016/j.soilbio.2004.02.009

[98] Nyamadzawo G, Wuta M, Nyamangara J, Smith JL, Rees RM. Nitrous oxide and methane emissions from cultivated seasonal wetland (dambo) soils with inorganic, organic and integrated nutrient management. Nutrient Cycling in Agroecosystems. 2014;**100**:161-175. DOI: 10.1007/s10705-014-9634-9

[99] Iqbal S, Xu J, Khan S, Worthy FR, Khan HZ, Nadir S, et al. Regenerative fertilization strategies for climate-smart agriculture: Consequences for greenhouse gas emissions from global drylands. Journal of Cleaner Production. 2023;**398**:136650. DOI: 10.1016/j.jclepro.2023.136650

[100] Malyan SK, Bhatia A, Kumar SS, Fagodiya RK, Pugazhendhi A, Duc PA. Mitigation of greenhouse gas intensity by supplementing with Azolla and moderating the dose of nitrogen fertilizer. Biocatalysis and Agricultural Biotechnology. 2019;**20**:101266. DOI: 10.1016/j.bcab.2019.101266

[101] Singh S, Singh DR, Velmurugan A, Jaisankar I, Swarnam TP. Coping with climatic uncertainties through improved production technologies in tropical Island conditions. In: Biodiversity and Climate Change Adaptation in Tropical Islands. London, United Kingdom: Elsevier; 2008. pp. 623-666. DOI: 10.1016/B978-0-12-813064-3.00023-5

[102] Kittas C, Katsoulas N, Bartzanas T, Bakker S. Greenhouse climate control and energy use. In: Baudoin W, Nono-Womdim R, Lutaladio N, Hodder A, editors. Good Agricultural Practices for Greenhouse Vegetable Crops. Rome: Food and Agriculture Organization of the United Nations; 2013. pp. 63-93

[103] Karanisa T, Achour Y, Ouammi A, Sayadi S. Smart greenhouses as the path towards precision agriculture in the food-energy and water nexus: Case study of Qatar. Environment Systems and Decisions. 2022;**42**:521-546. DOI: 10.1007/s10669-022-09862-2

[104] Butterbach-Bahl K, Kiese R, Liu C. Measurements of biosphere atmosphere exchange of CH 4 in terrestrial ecosystems. In: Rosenzweig AC, Ragsdale SW, editors. Methods in Enzymology. Vol. 495. London, United Kingdom: Academic Press; 2011. pp. 271-287. DOI: 10.1016/B978-0-12-386905-0.00018-8

[105] Heinemeyer A, Di Bene C, Lloyd AR, Tortorella D, Baxter R, Huntley B, et al. Soil respiration: Implications of the plant-soil continuum and respiration chamber collar-insertion depth on measurement and modelling of soil CO_2 efflux rates in three ecosystems. European Journal of Soil Science. 2011;**62**:82-94. DOI: 10.1111/j.1365-2389.2010.01331.x

[106] Hensen A, Groot TT, Van Den Bulk WCM, Vermeulen AT, Olesen JE, Schelde K. Dairy farm CH_4 and N_2O emissions, from one square metre to the

full farm scale. Agriculture, Ecosystems and Environment. 2006;**112**:146-152. DOI: 10.1016/j.agee.2005.08.014

[107] Pattey E, Edwards GC, Desjardins RL, Pennock DJ, Smith W, Grant B, et al. Tools for quantifying N_2O emissions from agroecosystems. Agricultural and Forest Meteorology. 2007;**142**:103-119. DOI: 10.1016/j.agrformet.2006.05.013

[108] Heinen M. Simplified denitrification models: Overview and properties. Geoderma. 2006;**133**:444-463. DOI: 10.1016/j.geoderma.2005.06.010

Section 2

Innovations for Economic and Environmental Improvement

Chapter 4

Innovative Greenhouse to Improve Economic and Environmental Conditions

Zainab Abdel Mo'ez Mansour Embaby

Abstract

Together with the World Bank and the Food and Agriculture Organization (FAO), a number of international organizations are promoting innovation in agricultural systems to combat natural disasters like extreme weather, drought, floods, rising sea levels, increased snowmelt, and changes in the amount and timing of water used for irrigation. The impacts of climate change on food security are undeniably significant, and they are expected to get worse over the coming years as a result of population growth, economic development, urbanization, and the recurrence of natural disasters. In today's agribusiness, particularly horticultural agribusinesses such as vegetables and decorative plants, climate-smart greenhouse is not a novel concept. In terms of GHG (greenhouse gas) emissions, CSA (Climate Smart Agriculture) can contribute. These days, climate-smart greenhouse (CSG) can actually connect adaptation and mitigation at all scales and helps farmers take the lead in combating climate change. The research on CSG emphasizes the need for innovative thinking to harmonize policy and practices in a way that is complementary. Additionally, CSG has to have a better grasp of how well-equipped the consultants or extension services are in each nation to assist with training farmers in climate-smart practices. Additionally, new financial tools are required to enable global, national, and local transformations.

Keywords: greenhouse, climate change, food security, adaptation, mitigation, innovative thinking, climate smart emissions, climate smart agriculture

1. Introduction

The world is currently dealing with a difficult, complex, but solvable set of issues as part of its ambitious attempt to achieve self-sufficiency in food production. Climate change modifies agricultural production and food systems, posing hazards of vulnerability and unpredictability to farmers and those who create policy. Planning for adaptation can take into account scientific data from both assessments of adaptable capability and estimates of climatic consequences, **Figure 1** ([1], pp. 8537-8362) clarified Impact approaches ([2], pp. 2775-2789; [3], pp. 607-610; [4], pp. 4422-4443). In view of analyzing global climate forcings and circulation models, they suggested that the main factors

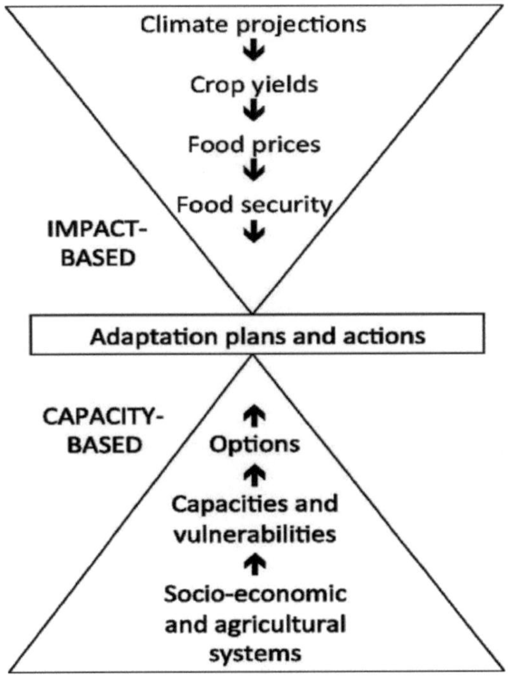

Figure 1.
Impact and capacity approaches to adaptation planning. Source: https://www.researchgate.net/figure/impact-and-capacity-approach.2013.

influencing crop yield are the connections between simulation and real-world adaptation to comprehend and predict climate change.

Climate change is harmful. The studies [5, 6] affirmed that climate change and variability (CCV) affect crop harvesting, including decreased rainy days, prolonged dry spell, sea-level rise, drought frequency and severity, heat stress, wind, pest, and disease outbreaks activities resulting in changes in rainfall patterns around the world with increasing flood. According to United Nations Environmental Protection Agency [7], climate forcing refers to a change in the Earth's energy balance, and a variety of natural and human variables can affect the Earth's energy balance and contribute to climate change. Burning fossil fuels, destroying forests, and preparing land for towns, roads, and farmland are all examples of human activity. It was concluded that all of these actions contribute to the atmospheric emissions of greenhouse. However, the Intergovernmental Panel on Climate Change [8] predicts that global warming will exceed the 1.5C upper limit this century, without rapid and significant cuts in greenhouse gas emissions. **Figure 2** clarifies adaptation plans and actions that keep global warming to 1.5 degrees Celsius with little to no overshoot ([8], pp. 13-14).

A chart (**Figure 2**) shows GHG emission reduction needed to keep 1.5 degrees C within reach. (IPCC AR6). Since fossil fuels are the primary source of GHG emissions and one of the causes of global warming, the phase-out of these fuels must be accelerated throughout society. Climate change impacts on agriculture will make it difficult to meet the key Sustainable Development Goals (SDGs) of ending hunger, achieving food security, and ensuring sustainable food production systems by 2030. In the longer term, facing the challenges to the quantity and quality of foods, urgent action is urgently needed to achieve food security. Agriculture is the affected sector of food security in all dimensions, especially food availability, through extreme weather events. On the other

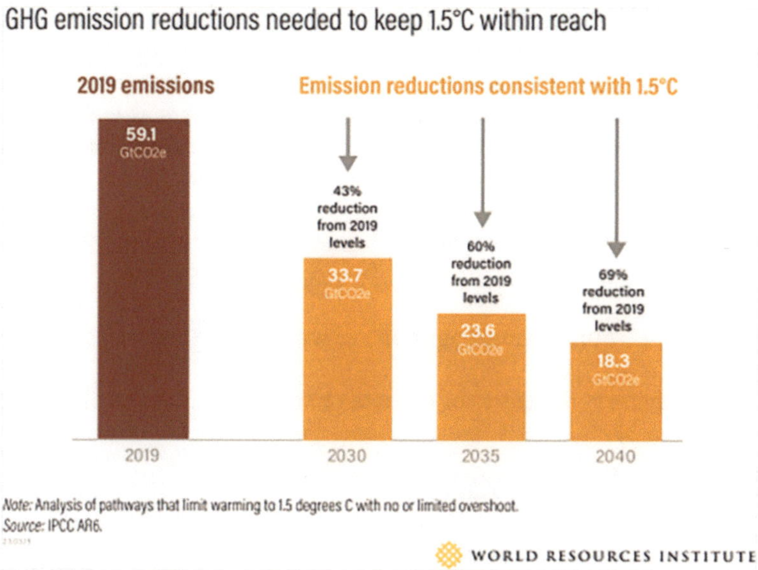

Figure 2.
GHG emission reductions consistent with 1.5°C from 2019 emissions to 2040 emissions. Source: [9].

hand, climate extremes are considered one of the challenges to the quantity and quality of foods people can access. Food Agriculture Organization ([10]; [11], pp. 521-546) released agriculture is a sector contributing both carbon emissions and capture uniquely susceptible to climate and extreme weather. In addition, agricultural innovations can combat climate change through both mitigation and adaptation (World Bank group [12]). To accommodate climatic conditions, agricultural activities will need to be modified to reduce greenhouse gas (GHG). Climate change is the only one of the major forces which will affect the future of agriculture. Others include population growth and increases in income as well as changes in human capital, knowledge, and infrastructure. Much of the changes in agriculture will stem from new innovations. The previous studies [13, 14] affirmed the role of Climate Smart Agriculture (CSA) in response to climate change. CSA plays a prominent role in facing increased demand for food. The CSA approach has been considered an essential mechanism for achieving the Sustainable Development Goals.

2. The significance of the chapter

With a share of 560 m3 of water per person, Egypt has become one of the most water-scarce nations in the world (United Nations International Children's Emergency Fund [15]. Additionally, Egypt may soon run out of water, with climate change being the primary cause. CSG contributes significantly to the community's revenue in rural areas, even in the absence of population growth and the race to enhance agricultural productivity. As a result, the emphasis of this analysis is on the significance of climate-smart greenhouse (CSG) as a cutting-edge remedy for food insecurity both globally and in Egypt.

3. Methodology

The current review study focused on numerous data found in English-language peer-reviewed papers worldwide with searches using terms relevant to CSA practices and CSA outcomes. The objective of this ongoing review is to give a first appraisal of the evidence for CSG as an innovative one contributing to improving economic and environmental conditions. This review highlighted Egypt, aiming to offer effective supporting information to decision-makers and policy makers as well as overall professionals and end-users in introducing new techniques, artificial intelligence, and communication infrastructure in agriculture sector. Then it focuses on:

- factors contributing to development of greenhouses and technologies worldwide (Section 1).
- overview of the climate-smart agriculture worldwide (Section 2).
- overview of the severe effects of climate change on the agriculture sector in Egypt (Section 3).
- CSA and GHG emissions (Section 4).
- economic and environmental benefits of application CSG (Section 5).
- application of CSG (Section 6).
- factors affecting traditional agriculture (Section 7).

3.1 Development of greenhouses and technologies worldwide

A number of significant factors, including population growth; urbanization; wealth development; changes in human capital, knowledge, and infrastructure; as well as climate change, have resulted in the introduction of novel characteristics to traditional agricultural farming methods [16]. The study conducted to release in Qatar to boost the local food and achieve its National Vision 2030, particularly the food security, environmental, and sustainability challenges, focused on differentiating innovations based on their forms, such as technological, managerial, and institutional innovations, in line with the economic growth hypothesis. It also clarified that technical innovation takes the form of new tools, mechanical innovations (like tractors), biological innovations (like seeds), chemical innovations (like fertilizers), better practices like Integrated Pest Management, enhanced pruning methods, and crop rotation serve as better practices' equivalents to managerial innovations, which are not physically represented in capital. Institutional innovations can refer to novel organizational structures, like cooperatives, and trading agreements, like futures markets and contract farming [17]. Due to the variety and irrationality of the effects of climate change, there are many different sorts of innovations. Sapkota et al. [18] propose that a way forward to address food security, climate change adaptation, and mitigation challenges faced by current agriculture is to widely promote suitable conservation agriculture (CA) practices by integrating them into national agriculture development strategies. The benefits of CA in terms of food security, climate change adaptation, and mitigation have been demonstrated in the Indo-Gangetic Plains (IGP) based on the findings of numerous farm and station trials. Due to greater accessibility and availability of food, there will be an increase in farm productivity and income for household food security. Similar improvements in crop yield; higher energy, water, and nutrient usage efficiency; as well as the least amount of heat stress show adaptability to climate change and unpredictability. It's still early to adopt and integrate new CSA technologies, like drones, and big data applications, including artificial intelligence and machine learning. Although multidisciplinary CSA research in Africa has advanced significantly, there is still a vacuum in the application of policies. Barasa et al. [19] made it clear that in order for the sub-Saharan region to achieve benefits from CSA, concrete steps must be taken to, among other things, encourage farmers to implement context-specific CSA technologies, make funds available to them, encourage investments, and create policy frameworks that support CSA.

3.2 Overview of effects of climate change on agriculture in Egypt

Along with others, [20] stated in light of the severe effects of climate change on the agricultural sector in Egypt, it was made clear that climate adaptation is a major national priority to preserve food security. In order to address the dispersion of funding schemes, it was further underlined that the Ministry of Environment should create a specific Climate Financing and Resource Mobilization Unit for adaptation in agriculture. International Monetary Fund (IMF) [21] clarified that as a result of climate change, the nations in the Middle East and Central Asia (ME&CA) have similar macroeconomic policy problems. The past economic effects of the region's main climate stressors—lower growth, shifting GDP and employment shares, and larger fiscal and external imbalances—will likely get worse with the predicted intensification of the region's climate stressors, especially where current weaknesses in climate resilience persist. Therefore, even under the assumption of mild global warming and ambitious global mitigation measures, regional policymakers must acknowledge that climate change would have an influence in the past three decades: variations in temperature and precipitation patterns have decreased per capita earnings and changed the sectoral composition of the economy, as econometric study claims. However, climate adaptation is an urgent priority for the region and requires significant additional spending and hence financing.

3.3 CSG and emissions reductions

The studies [6, 19] agreed upon empirically identifying factors that affect the intensity of participation in emission practices in Ghana and determining if adopting climate-smart agriculture practices decreases participation in emission practices. The study used inverse-probability-weighted regression adjustment (IPWRA) to achieve its goals, and empirical findings indicated that CSA can be applied as a method to lower GHG emissions from agricultural sources. Additionally, the government should take into account CSA technology installation as part of its policy. Also, it confirmed that the methodological approach is regarded as a robust one because it produces estimates that are nearly uniform across the IPWRA, the generalized Poisson model, and both Poisson and Poisson models. On the other hand, the studies affirmed that CSA increases profit, and minimizes vulnerability by reducing greenhouse gas emissions, by smart and advanced technological knowledge.

3.4 Economic and environmental benefits of application CSG

A case of Villages Around Songe-Bokwa Forest, Kilindi District, Tanzania [22] revealed that there was rainfall variability, shift in rainfall patterns, and increase in temperature in the study area. **Figure 3** shows impacts of climate change on household

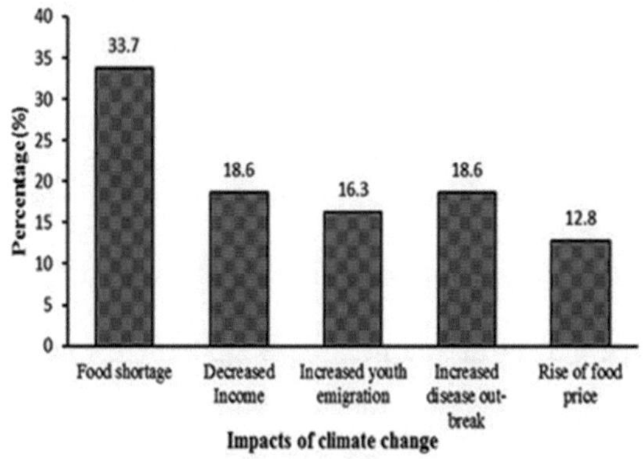

Figure 3.
Impacts of climate change on households. Source: The results of the study Nkumulwa & Pauline [22].

livelihoods, showing that 38.7, 18.6, and 12.8% of households perceived that climate change variability (CCV) resulted in food shortage, decreased income, increased disease outbreaks, youth emigration, and rise of food price.

A notable increase in crop harvest after farmers engaged in CSA was recorded in **Table 1**. The findings show farmers were food secured and gained more income through sales of their crops, and they used part of their income for paying school fees, buying production tools, supporting medical services, purchasing livestock, and paying for house construction. Consequently, CSA farmers became more resilient to negative climate effects. The study used random and purposive sampling designs to collect quantitative and qualitative data. Data in this study on the contribution of CSA to farmers livelihoods. Data in this study, on the contribution of CSA to farmers livelihoods, was subjected to analysis of variance(ANAVO) using the SPSS software package for Windows. Were more food secured and gained more income. In response to the decline in crop productivity and deforestation, the findings showed that farmers engaged in CSA practices such as agroforestry (i.e. agrisiliviculture), conservation agriculture, integrated nutrient management, and agronomic techniques such as cover crops, improved crop varieties, drought-resistant crops, intercropping, and crop rotation. Also, the production of crops after the introduction of CSA was higher than before the practice ($\alpha = 0.05$, df = 5, $p = 0.028$).

Crop type	No. of responses	Before CSA practice	After CSA practice
Maize	56	5.21	11.02
Beans	56	2.86	6.41
Pigeon pea	47	2.52	4.43
Tobacco	36	12.34	17.05
Mango	53	57.23	74.15
Cassava	42	41.22	64.82

CSA, climate-smart agriculture. Source: From the results of the study Nkumbula & Pauline [22].

Table 1.
Comparison of crop harvest per acre in a bag of 90 kg for climate-smart farmers before and after engaging in CSA interventions.

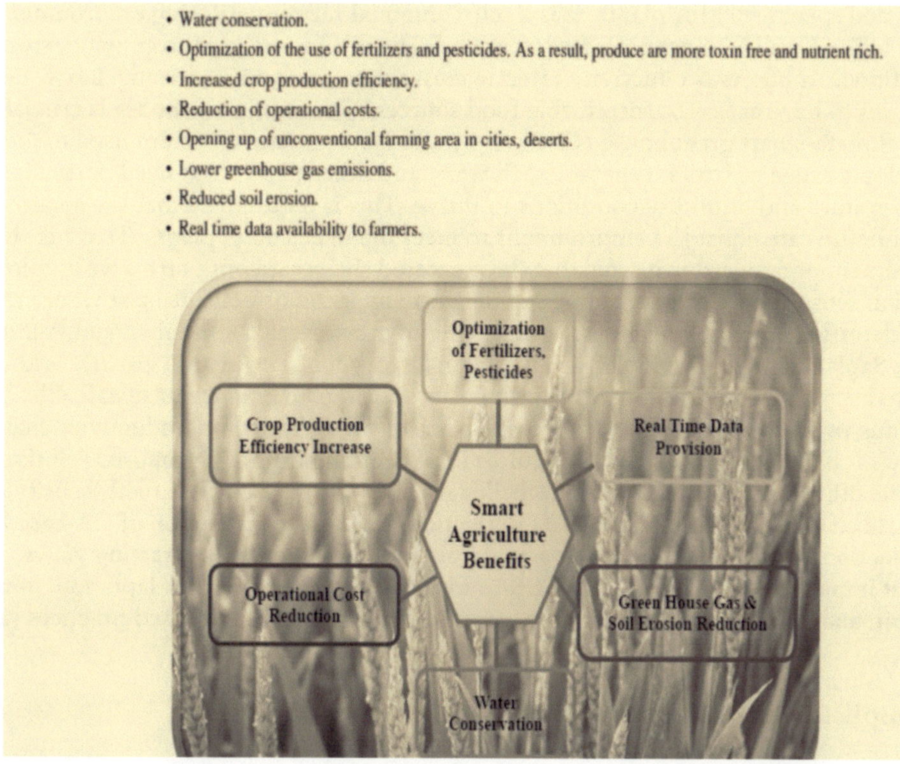

Figure 4.
Smart agriculture benefits over traditional agriculture. Source: https://www.researchgate.net/publication/357824463

Benefits of CSG include scheduling productions so as to maximize output, improve quality, and minimize waste. **Figure 4** shows the tremendous benefits of smart agriculture compared to traditional agriculture. ([23], pp. 1-45) concise the benefits of CSA as water conservation; optimization of the use of fertilizers and pesticides making products are more toxin-free and nutrient-rich. In addition to, the benefits include increased crop production efficiency; reduction of operational costs; opening up of unconventional farming area in cities, deserts; lower greenhouse gas emissions; reduced soil erosion; real time data availability to farmer. Also, production and distribution of food will be in economically efficient way as never before.

The automation of greenhouses has advanced significantly in recent years, largely due to environmental sensors that are essential to its programmed operation. In fact, modern sensor technologies integrated into smart greenhouse solutions are now frequently utilized to track the environment for crop growth. Using DHT 11 sensors to collect temperature and humidity data, it is possible to compare the conditions inside and outside a smart greenhouse for fruitful crops. While assuring effectiveness and sustainability, the integration of smart systems can decrease reliance on labor and boost profitability.

4. Application of climate-smart greenhouse and challenges

According to a survey, the majority of studies concurred on the key attributes of the smart greenhouse. It is [2, 16, 23] clarified a climate-controlled indoor space designed specifically for plants. It is a self-contained farm monitoring environment with IoT, AI, and ML technologies integrated. The farm is shielded from wind, storms, and floods. It boosts productivity effectiveness without requiring manual labor. For humans to have access to sustainable food sources, the smart greenhouse is crucial. The climate-smart greenhouse (CSG) application is a structural system used to develop a range of fruits, vegetables, flowers, and other plants that need particular temperature and humidity conditions to thrive. This is required so that the smart greenhouse can adjust the environment to meet the needs of its plants. To track the movement of dangerous insects that have entered the greenhouse farm, we employ a motion sensor. We can reduce insecticide waste by using insecticides just where they are identified, avoiding unnecessary spraying in other areas. Nkumulwa and Pauline, [22]; Sapkota et al. [18] highlighted the main factors: high population growth, and limited support from the government that drive farmers to practice unsustainable farming practices. Food security represents one of the agricultural productivity challenges. It is the greatest risks that requires implementing CSA as a proposed solution. On the other hand, to address triple challenges of present agriculture: food security, climate change adaptation, and GHG mitigation, wide-scale promotion of CA-based production system could be an important government strategy. Contrasting views about implementation indicate that CSA's focus on the "triple win" (adaptation, mitigation, and food security) needs to be assessed in terms of science-based practices [24].

5. Implications of climate-smart greenhouse research

CSG focuses especially on agriculture. It refers to an approach that sustainability increases productivity, enhances resilience (adaptation), reduces GHGs (mitigation) where possible, and enhances achievement of food security and development goals. CSA approach assessment is based on science-based practices. CSG can only be achieved in the long term after understanding and mitigating any challenges as the new paradigm shifts. Innovative thinking is required in order to reconcile policy and practices along complementary lines. CSG implementation also faces a better understanding of the capacity of extension services or consultants in each country to help train farmers on climate-smart practices. It is well-known that innovative technologies require specific extension support, sometimes not readily available. It's also important to comprehend the attitudes and behaviors of farmers regarding CSA activities. It all boils down to whether or not individual farmers are prepared to make the necessary adjustments or have the skills and knowledge to do so. Additionally, new financial tools are required to facilitate changes at all scales, including local, national, and global [20]. Along with the UNFCCC negotiations and the COP27 - Agriculture & Climate Change, new funding mechanisms are being developed for both climate change and agriculture. The recently concluded COP27 (November 23) of the United Nations Framework Convention on Climate Change (UNFCCC) offers a chance to start the shift to regenerative agriculture, under a whole food systems approach, which can bring various benefits for climate, health, resilience, biodiversity, and social justice. It concentrated on the need to see a 10-fold increase in climate finance to change agriculture and food systems for food and economic security by 2030. Innovative finance has a significant role to play in this.

Egypt performed a super job encouraging climate-smart agriculture to respond to the region's urgent agri-food and climate change requirements. It focused on the necessity of a 10-fold increase in climate finance to transform the food and agricultural systems for both economic and food security by 2030. Innovative finance can play a big part in this. Also, it can reduce food loss and waste and deal with the deterioration of irreplaceable natural like soil and water, and use ways to deal with heat, drought, and water scarcity under forecasted climate change scenarios. FAO Egypt at COP27 Hybrid Event, Sharm El-Sheikh (Egypt), 6–18 November, 2022. FAO Egypt at COP27 FAO in Egypt Food and Agriculture Organization of the United Nations.

6. Conclusions

In this chapter, we have seen how CSG is both a technical and a political concept that requires multidisciplinary work. We have also demonstrated how difficult it is to simultaneously implement the three pillars of CSG. We have discussed the main obstacles to CSA's adoption as well as its major policy and decision-making ramifications. Globally, rising temperature trends, an increase in the frequency of weather extremes, and an increase in seasonal variability have all been identified as new dangers to agriculture. Due to direct greenhouse gas emissions, agriculture has now been identified as one of the causes of climate change. Due to its potential involvement in GHG mitigation, agriculture is now starting to be seen as a way to combat climate change. A climate-smart greenhouse can aid in the creation of land-use plans that make the connectivity of adaptation and mitigation possible at all scales, thereby assisting farmers in taking the lead in the fight against climate change. The main takeaways from this chapter are as follows: (1) CSG meets sustainability, productivity, mitigation, food security, and development goals; (2) creative thinking is necessary; (3) a deeper comprehension of farmers' perspectives is also necessary; (4) additional financial instruments are required; and (5) a deeper comprehension is required of how well-equipped extension services or consultants are in each nation to assist in educating farmers about climate-smart practices. There is still room for improvement in policy implementation at the level of small farmers. To benefit from CSG, concrete steps must be taken to encourage farmers to use CSG technologies, provide them with the right funding, and encourage investment.

Acronyms and abbreviations

AI	artificial intelligence
ANOVA	analysis of variance
CA	conservative agriculture
CCV	climate change and variability
CSA	climate smart agriculture
CSG	climate smart greenhouse
EPA	United States environmental protection agency
GDP	gross domestic product
HGs	green house gases
GHG	green house gas
IMF	international monetary fund
IGP	Indo-Gangetic plain
IoT	internet of things
IPCC	intergovernmental panel on climate change

IPWRA	inverse-probability-weighted regression adjustment
ME&CA	middle east and central Asia
ML	machine learning
SDGs	sustainable development goals
UNCIF	United Nation children international fund
UNFCCC	United Nations framework convention on climate change

Author details

Zainab Abdel Mo'ez Mansour Embaby
Department of Environmental and Economic Assessment, Agriculture Research Center, Giza, Egypt

*Address all correspondence to: zainab.embaby@yahoo.com

References

[1] Vermeulen S et al. Addressing uncertainty in adaptation planning for agriculture. Proceedings of the National Academy of Science. 2013;**110**:8357-8362. DOI: 10.107/pnas.1219441110

[2] Challinor AJ, Ewart F, Arnold S, Simelton E, Fraser E. Crops and climate change: Progress, trends and challenges in simulating impacts and informing adaptation. Journal of Experimental Botany. 2009;**60**(10):2775-2789. DOI: 10.1093/jxb/rep062

[3] Lobell DB, Bure ME, Tebaidi C, Masterandrea MD, Naylor R. Prioritizing climate change adaptation for food security in 2030. Science. 2008;**319**(5865):507-610. DOI: 10.1126/science.1152339

[4] Nashwan MS, Shahid S. A novel framework for selecting general circulation model based on the spatial patterns of climate. International Journal of Climatology. 2020;**40**:442-4443. DOI: 10001002/joc.6463

[5] Fakan ST. Causes of climate change: Review article. Global Journal of Pure and Applied Sciences. 2020;**20**(2), Version 1:6-12

[6] Israel MA, Amikuzuno J, Abbeam GD. Assessing farmers contribution to greenhouse gas emission and the impact of adapting climate-smart agriculture on mitigation. Ecological Process. 2020; **9-51**:1-10

[7] EPA, 2014, United States Environmental Protection Agency Plan (EJ) 2014. USEPA. Available from: www.epa.gov/environmentaljustice/plan-ej-2014

[8] IPCC. Intergovernmental Panel on Climate Change. AR6 Synthesizes Report: Climate Change. IPCC. 2023. Available from: https://www.IPCC.ch/report/six

[9] World Resources Institute. 1o Big Finding from 2023 IPCC Report on Climate Change by Sophine Boem and Clea Schumer. Cover Image by Arinut Thailand/Shutterstock. World Resources Institute; 2023

[10] FAO. In: FAO, editor. FAO; IFAD; UNICF; WEP; WHO Serial. The State of Food Security and Nutrition in the World. 2022. DOI: 10.4060/cc0639en/report

[11] FAO. FAO Egypt at COP27 FAO in Egypt Food and Agriculture Organization of the United Nations. FAO; 2022

[12] World Bank Group. AADAPT: Agriculture Adaptation and Natural Resource Management, World Bank Group. 2023. Available from: https://www.worldbank.org/en/research/dime/brief/agriculture

[13] Paul M et al. A review of climate-smart agriculture research and applications in Africa. Agronomy. 2021; **11**(6):1255. DOI: 10.3390/agronomy 11061255

[14] Theodora K et al. Smart greenhouses as the path towards precision agriculture in the food-energy and water nexus: Case study of Qatar. Environment System and Decisions. 2022;**42**:521-546. DOI: 10.10071s 10669-022-0966-2

[15] UNICEF, United Nations International Children's Emergency Fund. Water Scarcity in Egypt, UNICEF, 2021. Available from: https://www.uncif.org/egypt/documents/water-scarcity

[16] Karanisa T et al. Smart greenhouse as the path towards precision agriculture in

the food-energy and water-nexus: Case study of Qatar. Environment Systems and Decisions. 2022;**42**:521-546

[17] Ziberman D et al. Innovation to response to climate change. Climate smart agriculture: Chapter. In: PART of the Natural Resource Management and Policy Book Series. Vol. 52. NRMP; 2018 First online: 21 October 2017

[18] Sapkota TB et al. Climate change adaptation, greenhouse gas mitigation and economic profitability of conservative agriculture. Some examples from cereal systems of Indo-Gangetic plans. Journal of Integrative Agriculture. 2015;**14**(8):1524-1533

[19] Barasa PM et al. A review of climate-smart agriculture research and applications in Africa. Agronomy. 2021; **11**(1255):1-26. Available from: https://www.mdpi.com/journal/agronomy

[20] Barakat F et al. From Policy to Implementation: Adaptation to the Impact of Climate Change on Agriculture on Egypt. Egypt: The American University in Cairo; 2022

[21] International Monetary Fund (IMF). Climate Adaptation for MF&CA. IMF; 2022

[22] Nkumulwa HO, Pauline NM. Role of climate-smart agriculture on enhancing farmers 'livelihoods and sustainable forest management: A case of livings around song-Bokwa Forest, Kilindi District, Tanzania. Frontiers in Sustainable Food Systems. 2021;**51**: 6714190. DOI: 10.3389/

[23] Alakananda M et al. Everything you wanted to know about smart agriculture. arXiv. 2021.04754vI[cs CY]. DOI: 10.48550

[24] Torquebiau E et al. Identifying climate-smart agriculture research needs. Cahiers Agriculture Cultures. 2018;27: 26001, DOI: 10.1051/cagri/2018010. Available from: www.cahieragricultures.fr

Chapter 5

A Review of Controlled Environment Agriculture (CEA) Vegetable Production in Africa with Emphasis on Tomatoes, Onions and Cabbage

Taiwo Bintu Ayinde, Charles Fredrick Nicholson and Benjamin Ahmed

Abstract

This chapter reviews the available information about performance indicators for controlled environment agriculture (CEA) and conventional production systems in Africa with an emphasis on those arising from tomatoes, onions and cabbage production. We identified a small number of studies that reported, yields per land area, costs, cumulative energy demand (CED), global warming potential (GWP) and water use for either CEA or field-based production systems. The available information does not allow robust comparisons of CEA and field-based production for any of these indicators, which suggests the need for expanded and improved crop-specific data collection from existing operations and the usefulness of alternative approaches such as economic engineering.

Keywords: controlled environment agriculture, economic analysis, GHG emissions, vegetable production, Africa

1. Introduction

More than 25% of the world's population suffers from micronutrient deficiencies and related health problems [1, 2]. Increased vegetable consumption has been proposed as a mechanism to reduce the prevalence of non-communicable diseases (NCD) in low and middle-income countries [1]. However, increasing vegetable consumption faces the challenge of increasing availability (production) at affordable costs [2, 3]. Total conventional (field-based) vegetable production increased in Africa from 2001 to 2021, with both tomatoes (*Solanum lycospersicum L.*) and onions (*Allium cepa L.*) production increasing by 59%, and cabbage (*Brassica oleracea var. capitata L.*) by 65% [4]. Field-based vegetable production in Africa is often practiced close to water supply points, in swampy areas or along the littoral band with easy access to water. In such areas, farmers operate within an informal economy and cultivate plots generally less than 1 hectare in size [3]. Production practices are characterized by the use of limited machinery and other inputs and hand-powered technology such as the use of cutlass, hoe and irrigation boxes [3]. African vegetable farmers generally have access to limited information about technical recommendations [4, 5]. This contributes to a wide range of negative impacts on the environment that include reduction in crop yield and subsequently income and revenue, biodiversity loss, deterioration of water catchments, declining plot sizes, land degradation and greenhouse gas (GHG) emissions [5, 6].

The increased demand for affordable and nutritious food in urban areas has resulted in more demand for land and high migration from rural to urban areas, with workers willing to carry out conventional vegetable farming. Supply chains for vegetables produced in the open fields of rural are usually informal with low levels of coordination [1]. Disruptions in international food supply chains due to COVID-19 and economic and political instability in the region have also compounded the inability to attain regional self-sufficiency in vegetable consumption.

Recent years have seen increased discussion about whether alternatives to field-based vegetable production such as controlled environment agriculture (CEA) provide a mechanism to increase supply in urban areas. CEA comprises multiple types of approaches at alternative scales, including the production of plants, fish, insects, or animals using in- (home production or indoor gardens), medium- (e.g., community gardens), or larger-scale commercial operations, e.g., rooftop greenhouses, plant factories (PF) or vertical farms (VF) often using hydroponics, aquaponics or aeroponics, and growth chambers. These technologies control to varying degrees environmental parameters such as humidity, light, temperature and CO_2 to create optimal growing conditions [6–16]. CEA technologies are classified according to the type of facility and growing systems [16]. Soil-based CEA systems use regular soil or compost as the plant growth medium and the predominant type of greenhouses in Africa [5, 7, 12–16]. In contrast, hydroponic systems are soilless culture in which solutions containing nutrients are applied directly to the roots of the plants. Aquaponic systems combine fish cultivation and hydroponics plant production. In aquaculture, microbial activity converts fish excreta into nutrients. The nutrient-rich wastewater is then pumped through the hydroponic area for use as the plant nutrient. The plants take up the nutrients and clean the water, which is then cycled back into the fish tank.

There has been significant rise in global CEA production for over the past two decades; the primary crops in CEA production focus on vegetables that form part of the local diet across all income groups (e.g., tomatoes, peppers, spinach, cabbage, and other leafy greens). Whether sold on local markets or to 'niche' markets such as hotels, the price is the same as for soil-grown produce [17]. The global–defined here as including vegetables that are important in diets across all income groups (e.g., tomatoes, peppers, spinach, cabbage, and other leafy greens)—has increased by ~80% from $ 0.4 billion in 2013 to $8.5 billion in 2022, and it is projected to reach about $20 billion in 2026. The main production regions in 2022 were North America ($1.4 billion), Europe ($1.4 billion) and Asia-Pacific ($1.3 billion). Africa and the rest of the world contributed about 14% of the global total, $665 million [17, 18]. CEA operations can often use fewer pesticides, land and water per unit of product [1], and reduce susceptibility to pests, diseases and adverse weather conditions. CEA has the potential to produce year-round yields of high-quality produce with yields as much as 20 times higher [7, 8, 17]. Depending on the technologies compared, CEA can use considerably less water and with the use of renewable energy sources could reduce GHG emissions per kg of product [1, 8]. CEA has received particular attention for production in urban areas where demand is large, but land is limited. CEA may play a beneficial role in the production of vegetables, for which Africa's per capita consumption is the lowest in the world although demand is increasing. Africa's vegetable imports amounted to US$ 1.9 billion in 2013. Most of the vegetables imported (tomato, lettuce, and onion) can be produced in the region [17].

In addition to the potential impacts on the supply of vegetables, peri-urban CEA development in low-income counties could provide opportunities for young people migrating to cities from rural areas who are in search of more profitable and less physically demanding work than traditional farming or because rural livelihoods are now less viable [16]. Growers in Kenya, Nigeria and India who provide training in hydroponics or aquaponics identified four common categories of people who are interested in starting commercial CEA ventures: young people looking to start their own business; conventional farmers wanting to try a new approach or boost insufficient income; people in other professions seeking additional income; and white-collar workers who are close to retirement.

Despite the advantages of higher yields and lower use of some inputs [8–11] some CEA production systems (e.g., plant factories or vertical farms) do not appear to be particularly "climate-smart" in the sense that they can have high production costs, and use more energy and emit more GHG per unit product [8–11]. CEA activity based primarily on soil-based greenhouses is growing in parts of Africa [16]. Egypt and South Africa have seen development of large-scale greenhouse projects in cooperation with Dutch businesses, greenhouses have a 'considerable' presence in Kenya, and a few cases in other countries exists (Nigeria, Namibia, and Somaliland). Efforts to promote increased production of fruits and vegetables with CEA on small farms in low-income countries have achieved limited success. The uptake of CEA technology particularly has been limited in low and middle-income countries, particularly in Africa, with high costs for installation and maintenance cited as constraints [1, 11, 12]. However, distance to market, having government support and access to social media, additional information are key positive determinants of awareness that would help inform the potential role for CEA in Africa (e.g., [19–22]). To date, there are few empirical studies that carry significant ecological impacts, food insecurity, nutrition-related problems, farmer livelihood challenges, and persistent food system-related inequities on CEA production in Africa [16, 17].

The objective of this chapter is to review the available evidence about the costs and selected environmental performance indicators for CEA and field-based irrigated production systems in Africa with an emphasis on tomatoes (20.7 million tonnes) and onions (15.50 million tonnes) for being some of the predominant vegetables grown in 2021 [4]. Literature search showed up no public data to demonstrate the economic and environmental viability of large-scale CEA production of cabbages and other crops of the cabbage family (4.00 million tonnes) [2, 17].

The evidence is derived from previous studies that reported information about the costs per unit, yields per land area, cumulative energy per unit, GHG emissions an global warming potential (GWP) and water use for CEA and conventional vegetable production. This will provide more information on the potential role for CEA production systems in Africa and highlight the priority needs for additional research.

2. Materials and methods

This review considers six performance indicators per unit of product: cost of output, yields per land area, cumulative energy, GHG and GWP and water use for three crops (cabbage, onions, and tomato) in Africa. The information derives from studies with a Life Cycle Inventory (LCI) and Life Cycle Assessment (LCA) perspectives. LCI and LCA are related in the sense of providing accounting of comprehensive and systematic documentation of the impacts, processes and material flows of production but LCI focuses on current operational inputs and impacts (like energy use in a greenhouse) whereas LCA includes the inputs and impacts of 'embedded inputs' (for example, the energy to manufacture and dispose of steel used in a greenhouse). Previous reviews (e.g., [23, 24]) have adopted a similar approach to synthesis of the available information.

We developed a database with information about the six performance indicators using two approaches. For the first approach, we identified literature published from January 2000 to January 2023 using Scopus and Web of Science with search terms 'life cycle inventory' AND ('greenhouse' OR 'CEA') AND ('tomatoes') and evaluated which studies had information on production systems in Africa. Fifty-four articles were obtained from the initial search for tomato production. An additional three studies were identified from the literature cited by the articles identified through the database searches. The second approach employed a search in Google scholar for 'Africa' AND ('onions' OR 'cabbage' OR 'tomatoes') AND 'irrigated' and 'CEA production'. This search resulted in seven studies on irrigated and CEA production, of which three [15, 16, 25] contained multiple data points (for either for different crops or production systems).

Although 64 studies were reviewed, the number of observations for which specific values of the six indicators per unit of the product were reported or could be calculated was considerably smaller. Many studies reported ranges of values for aggregations of multiple crops (e.g., [15, 25]), which we deemed insufficiently specific for this review. We thus excluded the observations derived from these studies. Our review identified four observations for unit costs per (2 for tomatoes from [26] and two for general vegetables [5, 15]), and two values for each of Cumulative Energy Demand (CED; MJ/kg [26]), Global Warming Potential (GWP; kg CO_2eq/kg [26]) and water use (lts/kg [7, 16, 26]).

One possible reason for the limited number of studies with information for specific crops is the predominance and importance of "mixed vegetable" production systems with multiple crops grown throughout the year. In some sense, it is the overall performance of these systems that is important for the farms producing them, which are often of smaller scale. Thus, studies of the overall performance metrics of costs, yields and water use [15, 26] are more common than studies reporting that information for individual crops. In addition, most studies did not report all metrics of costs, yields, CED, GWP and water use; production indicators were more commonly reported than estimates of energy or GWP. In some cases, we made conversions of available data to estimate appropriate metrics. For example, water use was sometimes reported in units of liters per m^2 per day, which required an estimate of the length of the growing season in addition to the conversion of yields to kg/m^2.

The functional unit for our analysis (a measurement that is normalized across all systems for comparative purposes) was 1 kg of product (multiple vegetables or tomato) grown each year. We undertook conversion calculations (e.g., total yields to yields/ha or total GHG to GHG/kg or total water per ha to water/kg product) that were specific to each study. Yields were estimated in kg/m^2 [7] were converted to hectare by multiplying by 10,000 kg ha^{-1}. We converted values in local currencies to USD using exchange rates at the time data were collected in previous studies.

Revenues ($/ha) were calculated based on product yield (kg/ha) and output price, when available. Total costs of production include both variable and field costs, converted to \$/kg when needed using information on yields per hectare and costs per ha. The value of water used was converted to water per kg using volume and yield factors in [26]. Cold greenhouse vegetable was 20.0 kg/m^3 yield/water use, where 1 m^3 was equivalent to 1000 liters.

Quantities of energy in the standard unit of energy were expressed based on the International System of Units (SI), the joule (symbol J), is equal to 3600 kilojoules or 3.6 MJ. It was converted to megajoules (MJ) based on specific energy density for fuel (36 MJ l_1 in [26]) and later MJ/kg and was adapted for this review.

3. Results and discussion

As noted above, the number of observations with sufficient information for inclusion in this review is small (**Table 1**). Thus, it is difficult to make direct comparisons of revenues and profitability between operators who are cultivating the same crop in different systems and countries. Some studies [15, 26] provide insights about the impacts on the profitability of alternative irrigation systems for vegetable products, but do not allow comparisons of field and CEA production. More evidence is needed to understand the economic feasibility of CEA vegetable production in low-income countries and the extent to which experience from high-income countries is relevant for low-income countries. That is, no analyzes comparable to [11] have been conducted yet for low- and middle-income countries. Differences in costs and profitability for CEA and conventionally-grown produce may narrow as the initial investment is amortized, productivity increases, and new, cost-effective technologies become available [21, 22, 25, 26] but it is difficult to predict when this might occur.

There is anecdotal evidence that hydroponic systems may be more economically viable for vegetable production, including in dry land climates due to their minimal water use [https://bicfarmsconcepts.com]. In these systems, inexpensive, locally-available materials were used as substrates [23, 25, 26]. Where electricity was expensive or its supply irregular, pumps that do not need to lift or spray water on to the roots, such as the gravity-driven Kratky or ebb and flow technologies may be

Products, production system	Production cost	Cumulative energy demand (CED; kWh/kg)	Global warming potential (GWP; kg CO_2eq/kg)	Water usage	Total
Tomatoes					
Greenhouse	1	1	1	1	4
Open field	1	1	1	1	4
Multiple vegetables					
Greenhouse	0	0	0	0	0
Open field	2	0	0	1	3
Total					
Greenhouse	1	1	1	1	4
Open field	3	1	1	2	7

Table 1.
Summary of observations from review.

preferred. Where there is sufficient water and reliable electricity, aquaponics can be a viable form because it has two outputs—vegetables and fish—that provide complementary sources of income [26].

There was some evidence that enclosed structures using shipping containers and re-purposed buildings can house financially, socially and environmentally viable. CEA operations, partly because they have enabled entrepreneurs to set up in built-up urban locations where there is no space for greenhouses [16]. The higher operating costs compared to greenhouses, due to the need for LED lighting and air conditioning, could be offset by reduced fuel costs to transport produce to market. The risk of losing crops due to occasional electricity outages can be less than the risk of losing crops in transport from rural areas to urban markets due to fuel shortages or absence of adequate cold storage. Another reason why completely enclosed structures could viable is that parameters can be set to provide optimum conditions year-round, enabling the higher running costs to be offset by higher, and more consistent yields [16].

We identified only two values from one specific study that provided estimates of energy consumption and GWP per kg [26] (**Table 2**). In this case, more energy (0.01/0.46 MJ/kg) are required during the dry season for irrigation than the seasonal systems for cold greenhouse and open field vegetables, respectively. The difference in the values of energy consumption is much less for the cold greenhouse than the open field because Beninese have no access to electricity for irrigation and use generators fueled with oil with a higher GWP [26]. This is in contrast to [11] who found that both energy and GWP were higher for heated greenhouses. The higher mineral and organic nitrogen fertilizer rate as well as irrigation efficiency were reported to have contributed to the difference of GWP due to both production of fertilizers and field emissions of 0.37 CO_2eq/kg in cold greenhouse and open field (0.11 kg CO_2eq/kg) vegetable production. The more water supplied, the higher the leaching rate and soil moisture content. The higher the maximal soil moisture, the higher the denitrification rate. The more nitrogen supplied, the more Nr emitted and soil pH that will increase the rate of volatilization. Overall, the nature and amount of energy consumed per volume of irrigation water applied were critical to the climate change potential.

The maintenance of water pumps would limit the quantities of energy consumed as well as the irrigation efficiency. Second, it could also enhance crop yields at the edge of rivers where soils present a greater water retention capacity, lowering the need for irrigation water. Better irrigation management taking soil properties and local climate (evapo-transpiration) into account could improve the water use efficiency and also reduce (water losses by drainage).

Water use per kg product was based on ranges of water use per m^2 per day for tomatoes [26] and multiple vegetables [15, 25] (**Table 3**). These estimates are for vegetable production in Ghana where water is conveyed in 15-liter watering cans to irrigate. The range of values is large, 153–840 lts/kg, but it is consistent with estimates

Parameters	Greenhouse	Open field
Yield (kg/ha)	119,808	41,582
CED; MJ/kg	0.01	0.46
GWP; kg CO_2eq/kg	0.37	0.11

Source: Perrin et al. [26].

Table 2.
Yield, cumulative energy demand and global warming potential for tomatoes, greenhouse and open field.

Parameters	Open field	
	Maximum water usage	Minimum water usage
Yield, kg/ha	29,440	20,000
Yield, kg/m^2	2.944	2.000
Water, lts/m^2/day	5.0	14.0
Growing days	90	120
Water, lts/m^2	450	1680
Water, lts/kg	152.9	840.0
Sources: Perrin et al. [26]; Obuobie et al. [15]; Drechsel and Keraita [25].		

Table 3.
Estimated water use for open field [tomatoes and multiple vegetables].

for field-based lettuce production in the USA of 201 lts/kg [11]. Both values are considerably larger than the 21 lts/kg reported by [11] for CEA leaf lettuce. The GWP of urban garden tomatoes in Benin were reported to be 4–23 times larger than the impacts of tomatoes grown in European cropping systems, due to low and variable crop yields (high fuel consumption for irrigation, large nutrient flows and use of insecticides [26]).

Given the very limited information on costs in most reviewed studies (e.g., [14–16, 18, 21]), improvements in data collection and reporting would be helpful to improve our understanding of the potential for CEA compared to field-based production systems. First, it is relevant to collect and report data for specific crops to facilitate comparison between production systems. CEA operations often focus on one or a few crops, so crop-specific data (i.e., not 'vegetables') is needed for adequate comparisons. Reporting of both yield per crop and yield per year when those are different would better represent total production for the purposes of calculating costs, revenue and input requirements per unit of product. CEA operations typically produce multiple crops per year, but this can also be true for field products (e.g., cabbage entries that have multiple cropping periods per year). Reasonably accurate cost data are also needed to make relevant comparisons. Future studies can usefully distinguish better between costs (both variable and fixed) and revenues. For example, some studies (e.g., [15, 16, 21]) report only price information (which can be used to calculate revenues) or total revenue information, but not cost data. Most of the studies reviewed reported no specific costs, either in aggregate (for a ha or for a cropping season) or per unit. It would also be helpful if additional disaggregation of cost categories (especially for energy inputs like fuel and electricity) were reported because they would facilitate improved estimates of environmental impact. In general, it would be helpful if future studies were also more comprehensive, reporting information on all performance metrics we considered: yields, costs, and input use for energy and water.

More studies of CEA are needed for Africa, especially for sub-Saharan Africa. This can include both less technologically advanced systems (e.g., greenhouses without full temperature or humidity control), and more advanced (and expensive) systems such as greenhouses with more environmental controls—and similar systems such as plant factories and vertical farms. Generally, greenhouses and polytunnel structures were readily obtained locally [16], except in Nigeria, where greenhouses are not yet popular (they are imported and relatively expensive [6, 16, 20–23, 25–28]). Variations the combine different characteristics may be relevant. For example, hydroponic units have been installed outside of greenhouses in Nigeria, influenced by crop varieties and space constraints. In Kenya, suppliers offer hydroponic units that can be installed in a variety of enclosed or open settings, including a small unit that can be mounted on the wall of a building for those with no land.

4. Conclusions

The uptake of CEA production technology has been limited in low and middle-income countries, particularly Africa. This means that we have very limited information to evaluate the potential of CEA and to make comparisons to field production. That is, until we have more examples—and data—to evaluate it will be difficult to understand the potential role for, and impacts of, different types of CEA production systems in the region. One requirement is expanded and improved data collection from existing operations. Improved data collection would include a broad range of relevant indicators collected in a consistent manner across farms and studies. Another approach is to develop more 'synthetic' approaches based on economic engineering approaches used for other food production technologies [28]. This approach can suggest the conditions under which CEA operations may be successful even in locations where they do not currently exist. The long-term success and economic viability of CEA in Africa will also depend on future trends in consumer preferences and market demand for vegetables (e.g., [25, 28]), so studies on the costs would be complemented by consumer preference studies. Together, this information will inform decisions about private investors and governments.

Authors contributions

All authors contributed to the study conception and design. Material preparation, data collection and analysis were performed by [Taiwo Ayinde], [Charles Fredrick Nicholson] and [Benjamin Ahmed]. The first draft of the manuscript was written by [Taiwo Ayinde] and all authors commented on previous versions of the manuscript. All authors read and approved the final manuscript.

Conflict of interest

The authors declare no conflict of interest.

Ethical approval

All due consents have been sought.

Consent to participate

All due consents have been sought.

Consent to publish

All due consents have been sought.

Author details

Taiwo Bintu Ayinde[1]*, Charles Fredrick Nicholson[2] and Benjamin Ahmed[3]

1 Samaru College of Agriculture, Ahmadu Bello University (ABU), Zaria, Nigeria

2 Departments of Animal and Dairy Sciences and Agricultural and Applied Economics, University of Wisconsin-Madison, Madison, Wisconsin, USA

3 Department of Agricultural Economics, Ahmadu Bello University (ABU), Zaria, Nigeria

*Address all correspondence to: taiyeayinde2006@yahoo.com

References

[1] Pinstrup-Andersen P. Is it time to take vertical indoor farming seriously? Global Food Security. 2018;**17**(September 2017):233-235. DOI: 10.1016/j.gfs.2017.09.002

[2] Nicholson CF, Monterrosa E. Application of participatory systems modeling to identify intervention priorities for vegetable consumption in Nairobi Kenya. In: Revision for Public Health Nutrition. 2023

[3] Mondédji AD, Silvie P, Nyamador WS, Martin P, Agboyi LK, Amévoin K, et al. Cabbage production in West Africa and IPM with a focus on plant-based extracts and a complementary worldwide vision. Plants. 2021;**10**(3):1-36. DOI: 10.3390/plants10030529

[4] FAOSTAT. Crops and Livestock Products [Internet]. 2021. Available from: https://www.fao.org/faostat/en/#data/QCL [Accessed: May 4, 2023]

[5] Robert K, Kara HA, Adamu MM, Muhammad B. Analysis of dry season vegetable production among Kiri dam users in Adamawa state, Nigeria. Journal of Agripreneurship and Sustainable Development (JASD). 2021;**4**(4):173-179

[6] Dorr E, Sanyé-mengual E, Gabrielle B, Grard BJ, Aubry C. Proper selection of substrates and crops enhances the sustainability of Paris rooftop garden. Agronomy for Sustainable Development. 2017;**37**:51. DOI: 10.1007/s13593-017-0459-1

[7] Wachira JM, Mshenga PM, Saidi M. Comparison of the profitability of small-scale greenhouse and open-field tomato production systems in Nakuru-North district, Kenya. Asian Journal of Agricultural Sciences. 2014;**6**(2):54-61

[8] Dorais M, Antón A, Montero JI, Torrellas M. Environmental assessment of demarcated bed-grown organic greenhouse tomatoes using renewable energy. Acta Horticulturae. 2014;**1041**:291-298. DOI: 10.17660/ActaHortic.2014.1041.35

[9] Fan Y, Luo Z, Hao X, Li S, Kang S. Potential pathways to reduce environmental impact in a greenhouse tomato production: Life cycle assessment for different irrigation and fertilization treatments. Scientia Horticulturae. 2022;**305**(March):111411. DOI: 10.1016/j.scienta.2022.111411

[10] Fisher S. A Case Study of Urban Agriculture: A Life Cycle Assessment of Vegetable Production. Faculty of the Graduate School of the University of Colorado, ProQuest LLC; 2014

[11] Nicholson CF, Harbick K, Gómez MI, Mattson NS. An economic and environmental comparison of conventional and controlled environment agriculture (CEA) supply chains for leaf lettuce to US cities. In: Aktas E, Bourlakis M, editors. Food Supply Chains in Cities. Cham: Palgrave Macmillan; 2020. DOI: 10.1007/978-3-030-34065-0_2

[12] Aboaba KO, Sanusi RA, Akamo AA, Bello B. Double hurdle approach to consumer awareness, perception of, and willingness to pay for greenhouse vegetables. International Journal of Vegetable Science. 2022;**28**(1):15-24. DOI: 10.1080/19315260.2020.1819929

[13] Adams F, Etuah S, Appiah GB, Aidoo R, Osei J, Nyekyeyel J, et al. Do consumer opinions matter? Consumer perception and purchasing decisions of greenhouse vegetables in

Ghana. Journal of International Food & Agribusiness Marketing. 2022. DOI: 10.1080/08974438.2022.2145536

[14] Oyediran W, Omoare AM, Shobowale AA, Onabajo AO. Effect of socio-economic characteristics of greenhouse farmers on vegetable production in Ogun state, Nigeria. Sustainability, Agri, Food and Environmental Research. 2020;**8**(1):76-86. DOI: 10.7770/safer-v0n0-art1593

[15] Obuobie E, Keraita B, Danso G, Amoah P, Cofie OO, Raschid-sally L, et al. Irrigated Urban Vegetable Production in Ghana Production in Ghana: IWMI-RUAF-CPWF. Accra, Ghana: IWMI; 2006

[16] Halliday J, von Kaufmann R, Herath KV. An Assessment of Controlled Environment Agriculture (CEA) in Low- and Lower-Middle Income Countries in Asia and Africa, and Its Potential Contribution to Sustainable Development. Colombo, Sri Lanka: Commission on Sustainable Agriculture Intensification. CGIAR Research Program on Water, Land and Ecosystems (WLE); 2021. 86p

[17] de Janvry A, Sadoulet E. Agriculture for development. Development Economics. 2nd ed. Issue No 34. 2021. pp. 448-476. DOI: 10.4324/9781003024545-19

[18] Asongwe GA, Yerima PKB, Tening AS. Vegetable production and the livelihood of farmers in Bamenda municipality, original research article vegetable production and the livelihood of farmers in Bamenda municipality, Cameroon. International Journal of Current Microbiology and Applied Science. 2014;**3**(12):682-700

[19] Paucek I, Durante E, Pennisi G, Quaini S, Gianquinto G, Orsini F. A methodological tool for sustainability and feasibility assessment of indoor vertical farming with artificial lighting in Africa. Scientifc Reports. 2023;**13**:2109. DOI: 10.1038/s41598-023-29027-8

[20] Jones OE, Tham-Agyekum EK, Ankuyi BF, Ankrah DA, Akaba S, Shafiwu AB, et al. Mobile agricultural extension delivery and climate-smart agricultural practices in a time of a pandemic: Evidence from southern Ghana. Environmental and Sustainability Indicators. 2023;**19**:100274

[21] Kabiru F, Nina C, Kosgei B. Feasibility Study of the Best CEA System for School Feeding in Mukuru Informal Settlement, Nairobi. Berlin: TMG Research; 2023

[22] Gómez C, Currey CJ, Dickson RW, Kim HJ, Hernández R, Sabeh NC, et al. Controlled environment food production for urban agriculture. HortScience. 2019;**54**(9):1448-1458. DOI: 10.21273/HORTSCI14073-19

[23] Verteramo L, Nicholson CF, Gómez MI. A meta-analysis of life cycle assessments for tomato, lettuce and strawberry supply chains. Revision Submitted to Journal of Cleaner Production. 2023

[24] Li J, González W, Monterrosa E, Gómez MI, Nicholson CF. Choice experiments and value-chain modeling of attribute improvements to increase vegetable consumption in Kenya. Submitted to Food Policy. 2023

[25] Drechsel P, Keraita B, editors. Irrigated Urban Vegetable Production in Ghana: Characteristics, Benefits and Risk Mitigation. 2nd ed. Colombo, Sri Lanka: International Water Management Institute (IWMI); 2014. DOI: 10.5337/2014.219

[26] Perrin A, Basset-Mens C, Huat J, Benoit G. The variability of field emissions is critical to assessing the environmental impacts of vegetables: A Benin case-study. Journal of Cleaner Production. 2017;**153**:104-113. DOI: 10.1016/j.jclepro.2017.03.159

[27] BIC Farm Concepts. BIC Farm Concepts. 2023. Available from: https://bicfarmsconcepts.com

[28] Risner D, Li F, Fell JS, Pace SA, Siegel JB, Tagkopoulos I, et al. Preliminary techno-economic assessment of animal cell-based meat. Food. 2021;**10**(3):10010003. DOI: 10.3390/foods10010003

Chapter 6

Simulation of a Novel Cooling System for a Closed Greenhouse

Geordie Zapalac

Abstract

A simulation of a cooling system for a closed greenhouse is described. The cooling system relies upon cool ambient temperatures during the night and morning to discharge heat accumulated within the greenhouse during the day. Radiative heat into the greenhouse is transferred to a large reservoir of water inside the greenhouse using an unpressurized droplet system. During the night and morning the accumulated reservoir heat is discharged to ambient air using the same droplet system to transfer reservoir heat into a restricted volume of air above the reservoir, while simultaneously circulating the heated air through an air-to-air heat exchanger comprised of thin-walled plastic tubes.

Keywords: closed greenhouse, simulation, convective cooling, heat exchanger, water savings

1. Introduction

Climate change is creating a crisis for food security. Grain yields are threatened by elevated temperatures and drought stress that shorten the grain filling period and impair starch biosynthesis [1]. Increasing irrigation to counter high temperatures and drought is a progressively less likely option because climate change impacts other aspects of the hydrological cycle in addition to precipitation, including glaciers, river flows, and aquifer replenishment, increasing competition to agriculture for freshwater resources required for wild ecosystems, consumption and sanitation, industry, and cooling [2]. Climate change also reduces arable land by desertification of drylands [3], soil erosion from extreme precipitation events [4], and saltwater intrusion into river deltas [5].

Mitigating climate change will require removing gigatonnes of CO_2 annually from the atmosphere [6]. It is generally assumed that captured CO_2 will be liquified under pressure and geologically sequestered. For schemes where CO_2 is not mineralized underground and where no fluid is produced from the well, it has been argued that the increase in well pressure precludes underground sequestration of CO_2 at scales required to mitigate climate change [7]. Storage in saline formations would be possible by simultaneously producing brine to relieve the pressure, but this would require desalinating the produced brine and pumping the highly concentrated waste brine back into the formation for disposal [8]. Storage of CO_2 in oil wells during enhanced oil recovery (EOR) is possible because oil is produced to relieve the pressure. The CO_2 sequestered by EOR could be managed to exceed the CO_2 emitted by combusting the produced oil [9].

Use of captured CO_2 to enhance greenhouse yields provides a potentially profitable route of sequestration into biomass such as biochar, woody products, or humus, and CO_2 could be provided at ambient pressure from the output stream of the CO_2 capture facility. The greenhouses must be closed or unventilated so that the CO_2 is confined until it is consumed by the plants. Water is also conserved and recycled because it is confined within the greenhouse with the CO_2. Closed greenhouses can enhance yields of C_3 crops by maintaining a high concentration of CO_2 during the afternoon period of maximum photosynthesis, when other greenhouses are normally ventilated for cooling. High CO_2 concentrations might increase the temperature for optimal photosynthesis [10–14], increasing the yield while reducing the cooling load for the greenhouse.

Cooling a closed greenhouse generally requires much more energy than evaporatively cooling a ventilated greenhouse. However, renewable energy and energy storage costs are falling while freshwater resources are diminishing, and CO_2 will need to be sequestered at scale. Therefore an important engineering challenge for addressing food security and CO_2 sequestration in a changing climate is the problem of economically cooling a closed greenhouse.

Different solutions to the closed greenhouse cooling problem have been prototyped and commercialized in the past. Closed greenhouses sited in northern climates have used borehole heat exchangers to access cold ground temperatures that are recharged to low temperatures during the winter [15]. Closed greenhouses have been sited over aquifers to access seasonal storage of cold water temperatures [16]. The closed greenhouse cooling system for the Watergy prototype operated on a diurnal cycle using a water-to-air heat exchanger that accessed a reservoir of water outside the greenhouse that was cooled by low ambient nighttime temperatures [17, 18]. The Novarbo Oy Company commercialized a closed greenhouse that cools and dehumidifies the greenhouse air using water droplets, returning the water to an outside reservoir that is cooled with a heat pump [19, 20].

The closed greenhouse design described in this report relies upon cool ambient nighttime and morning temperatures to discharge heat removed from the greenhouse air during the day and stored in a reservoir of water inside the greenhouse [21]. During the day heat entering the greenhouse from solar radiation is transferred to a reservoir of water by an unpressurized droplet system. The novelty of the proposed design is the method of discharging the accumulated reservoir heat to the ambient air. During the night and morning heat in the reservoir is discharged to ambient air by using the same droplet system to transfer reservoir heat into a restricted volume of air above the reservoir, while simultaneously circulating the heated air through an air-to-air heat exchanger composed of thin-walled plastic tubes. This design avoids the cost and maintenance of a chiller as well as the supported weight and risk of leaks using a water-to-air heat exchanger. The design requires a climate with a low minimum ambient temperature during the morning, preferably 13°C or less. Because the greenhouse is convectively cooled, it could be sited equally well in deserts and in high humidity climates.

The greenhouse cooling simulation described here uses a "well-stirred" energy balance model where the temperature and humidity are assumed to be uniform throughout the greenhouse volume. Heat and mass transfers are modeled with correlation formulas developed for forced convection applications that are based upon dimensionless Nusselt numbers. These formulas have an accuracy of about 20% [22].

Experiments were performed on small components of the cooling system: a single plastic heat exchanger tube and droplet dispensers [21]. These experiments confirmed the air-to-air heat transfer properties of the plastic tube, and informed the simulation on the effective droplet surface temperature for simulating the heat and mass transfer to a falling droplet. The droplet surface temperature is modeled using a linear combination of the bulk temperature in the droplet and the surrounding air temperature [21].

The simulation advances in time steps of $\Delta t = 90$ seconds for a 24-hour period. More refined time steps are required to model the reservoir and heat exchanger system during the heat discharge period. The 24-hour simulation is repeated iteratively until the total heat within the greenhouse at the end of the 24-hour cycle is within 10 kJ of the total heat at the beginning of the cycle.

2. Components and operation of the cooling system

Figure 1 shows a schematic of four greenhouse modules taken from an array of closed greenhouses, where the cover is not shown for one of the modules in the drawing. The green cultivated regions in this example have an area of 20 m by 50 m or 0.1 ha. Adjacent to each cultivated region is a reservoir of water with a depth d_r of 1 m and an area A_r of 10 m by 50 m. Above each reservoir is a restricted volume of air or "tunnel" that is optionally open to a bank of plastic tubes located above the tunnel that serves as an air-to-air heat exchanger. The individual heat exchanger tubes are not shown.

During the day the reservoir water is cool, the sides of the tunnels are open to the cultivated regions in the greenhouses on either side, and the ends of the tunnels are

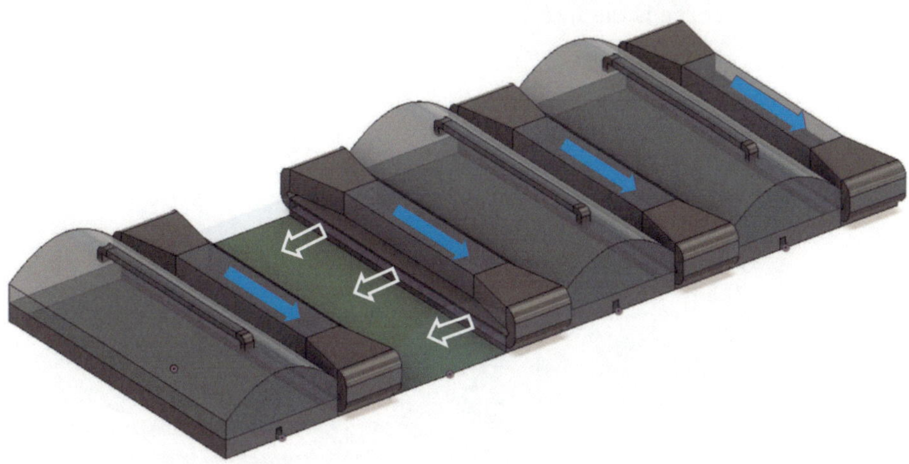

Figure 1.
Four modules from a greenhouse array that share a common volume of air. One of the modules has the cover removed from the drawing. White outlined arrows show the direction of airflow during the day, when warm greenhouse air is circulated through cool reservoir droplets to transfer heat and water vapor from the air into the reservoirs. Blue arrows show the direction of airflow during the night through the heat exchanger tube bank, when cool tunnel air is circulated through warm reservoir droplets to transfer reservoir heat into the air-to-air heat exchanger.

closed off from the C-shaped ducts that lead to the outside heat exchanger. 1.6 m above the reservoir surface, below the roof of the tunnel, there are trays with a pattern of short tubes that comprise the droplet dispensers.

Cool reservoir water pumped into the trays returns to the reservoir as 1.5 mm diameter droplets that exchange heat and water vapor with the air above the reservoir. Warm greenhouse air is circulated by fans across the width of the tunnel and through the falling droplets in the direction of the white outlined arrows, shown for one the modules, transferring both heat and water vapor to the reservoir to cool and dehumidify the air. A complete greenhouse array would be configured to return the airflow through a similar string of greenhouses in the opposite direction so that the airflow is cycled through all the greenhouse modules.

During the night the reservoir water is warm, the sides of the tunnel are closed, and the ends of the tunnel are opened to the ducts that lead to the heat exchanger tube bank. The same droplet system is activated and fans move saturated air down the length of the tunnel through the falling droplets, and then through the heat exchanger tube bank in the opposite direction, shown by blue arrows in **Figure 1**, where the forced convection of greenhouse air transfers both sensible and latent heat to the ambient air. Outside fans also pull a crossflow of cool ambient air through the tube bank.

The tube bank is comprised of 900 PETG tubes that are 30 m in length and in contact with the ambient air. Each tube has an outer diameter D_1 of 10 cm and a wall thickness of 0.5 mm. The tubes are arranged in a staggered configuration of 30 rows with 30 tubes per row, with a pitch of a =2 within a row and a pitch of b =1.25 in the vertical direction. The tubes may be slightly angled to allow condensed water to drain back into the reservoir. **Figure 2** is a closer view of one end of the cooling system showing the surface of the reservoir, the top of the tunnel where the droplet dispensers are located, the duct leading to the heat exchanger, and the region occupied by the heat exchanger tube bank.

Figure 3 shows the solar radiation into the 1000 m^2 greenhouse in the simulation. The integral of this curve is the heat load that must be transferred during the day to

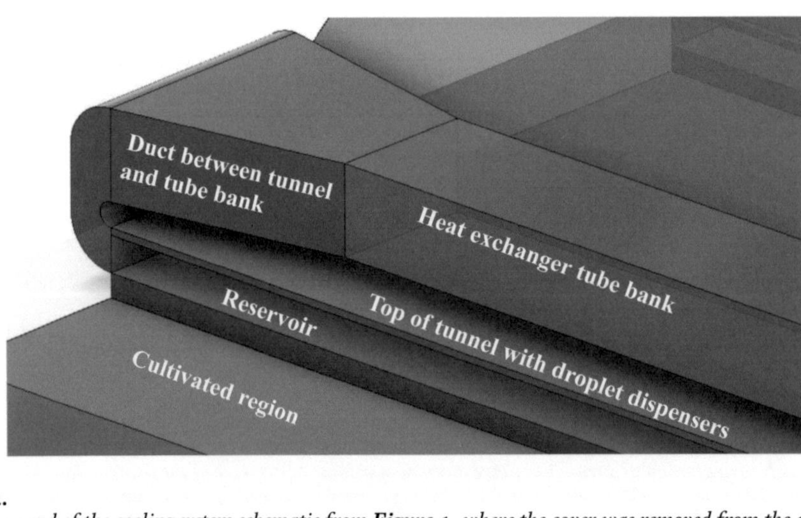

Figure 2.
*View of one end of the cooling system schematic from **Figure 1**, where the cover was removed from the drawing. Adjacent to the cultivated region is the primary heat exchanger system comprising a reservoir of water, a tunnel with droplet dispensers to transfer heat to or from the reservoir, and a C-shaped duct that leads to a bank of 900 thin-walled plastic tubes.*

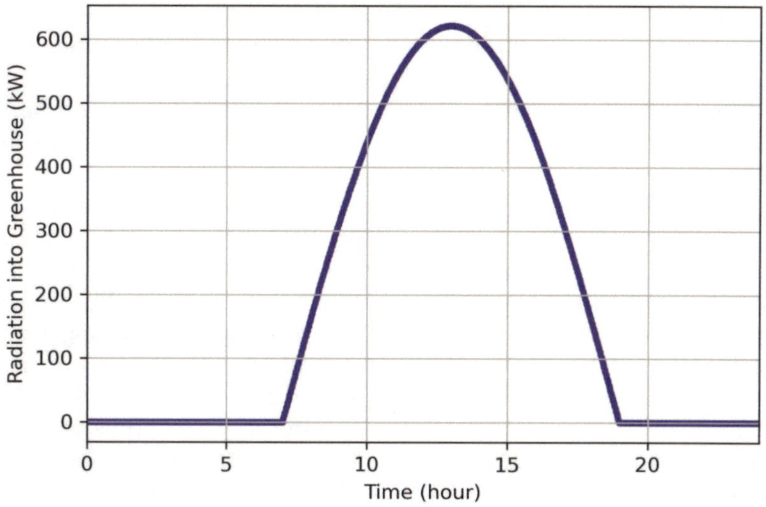

Figure 3.
Model of the daily radiative heat load into the 1000 m² greenhouse.

the reservoir or to the soil and plants, or conducted to the ambient air through the cover. If the integrated outside radiation is 9.5 kWh m^{-2} day^{-1}, then the transmission of the greenhouse cover must reduce the incoming insolation by 50% to achieve the radiation heat load shown in **Figure 3**. During the day the reservoir water temperature is roughly 15°C cooler than the greenhouse air. The reservoir droplet dispensers are activated to flow 100 L s^{-1} of droplets to transfer both latent and sensible heat into the reservoir. During the night the reservoir droplet dispensers flow 450 L s^{-1} to transfer heat from the reservoir to the air.

Figure 4 is a plot of temperature versus distance through 50 m in the tunnel, shown to the left of the vertical dashed line, and through 30 m of the heat exchanger tubes, shown to the right of the dashed line. It is assumed that the air temperature

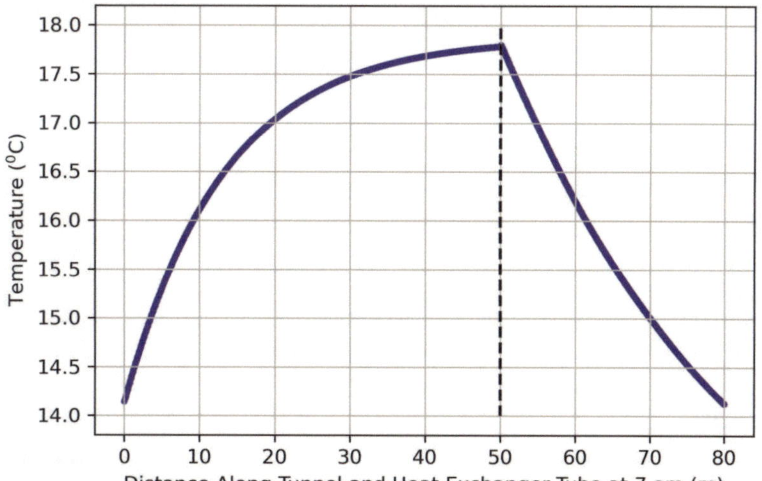

Figure 4.
Air temperature versus distance in the heat exchanger system. During the first 50 m the temperature rises in the tunnel as reservoir heat is transferred by warm reservoir droplets to the air. During the last 30 m the air cools as it passes through the heat exchanger and transfers heat originally stored in the reservoir into the cool ambient air.

does not change within the ducts; the distance through the ducts is not shown in **Figure 4**. As cooled air moves through the warm reservoir droplets down the length of the tunnel the temperature increases and the humidity remains saturated. When the air moves through the heat exchanger the temperature falls as reservoir heat is transferred to the ambient air. Water condenses inside the tubes and eventually drains back into the reservoir.

During the night heat is conducted through the walls and roof of the cultivated regions and the air temperature falls, increasing the relative humidity. A second air-to-air heat exchanger, denoted as the condenser, is positioned on the apex of each greenhouse roof as shown in **Figure 2** to reduce the relative humidity at night. The condenser has the same tube arrangement as the primary heat exchanger over the tunnel, but the PETG tubes are smaller, with an outside diameter of 2.5 cm and a wall thickness of 0.3 mm. Indoor fans pull air through the condenser tubes, while outside fans pull a crossflow of cool ambient air through the condenser tube bank.

Water is completely recycled within the closed greenhouse. All the water that evaporates into the air, including water transpired by the plants, eventually condenses on the falling reservoir droplets or on the inner surfaces of the heat exchanger tubes and returns to the reservoir.

3. Simulation

3.1 Daytime cooling

During the day radiative heat must be transferred to the reservoir by the droplet system at approximately the same rate that it enters the greenhouse. The plants also transpire and the additional water vapor must be removed by the droplet system to maintain an acceptable vapor pressure deficit (VPD) for the plants. **Figure 5** shows the model for plant transpiration over a 24-hour cycle, assuming a transpiration rate of 3 L m^{-2} day^{-1} in the cultivated region. The simulated transpiration rate at any

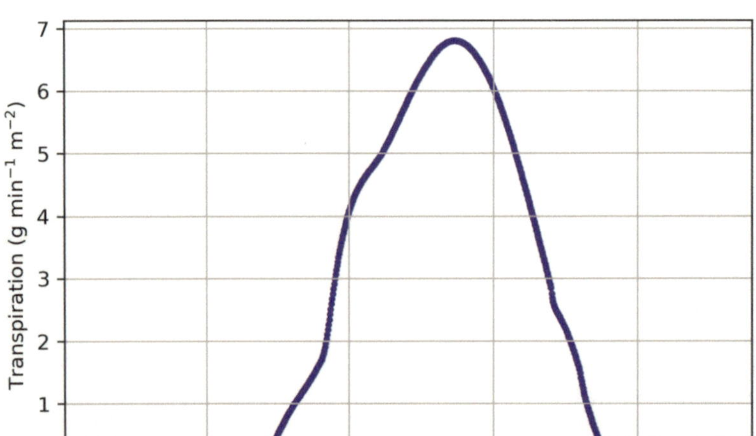

Figure 5.
Transpiration rate from plants cultivated in the greenhouse.

given time depends upon the insolation and VPD. Because the VPD depends upon the solution for the greenhouse temperature and humidity, the time dependence of the transpiration is iterated together with the simulated temperature and humidity to provide a self-consistent result after the simulation converges [21].

During a time step Δt the reservoir droplets absorb the heat ΔQ_r and the water vapor mass ΔM_r from the greenhouse air. These are signed quantities: ΔQ_r and ΔM_r are positive when heat and water vapor are transferred from the greenhouse air to the reservoir droplets, and negative when heat and water vapor are transferred from the reservoir droplets to the greenhouse air. During Δt the reservoir temperature increases by:

$$\Delta T_r = \frac{\Delta Q_r}{A_r d_r \rho_w C_{pw}} \tag{1}$$

where ρ_w and C_{pw} are the density and specific heat capacity of water. The mixing ratio X_a (grams of water per grams of dry air) for the total greenhouse air volume V_a changes by:

$$\Delta X_a = \frac{\Delta M_{Tr} - \Delta M_r}{V_a \rho_a} \tag{2}$$

where ΔM_{Tr} is the mass of water vapor transpired by the plants during Δt. During Δt the radiative heat ΔQ_S enters the greenhouse, and conductive heat $\Delta Q_C > 0$ (< 0) also enters (leaves) the greenhouse through the cover over the cultivated region. The greenhouse air temperature T_a changes by:

$$\Delta T_a = \frac{\Delta Q_S + \Delta Q_C - H_v \Delta M_{Tr} - (\Delta Q_r - H_v \Delta M_r)}{V_a \rho_a C_{pa} + m_s C_{ps}} \tag{3}$$

where H_v is the enthalpy of vaporization for water, C_{pa} is the specific heat capacity of air, m_s is the mass of the soil in the cultivated region (g), and $C_{ps}$is the specific heat capacity of the soil (0.92 J °C^{-1} g^{-1}). The simulation assumes that heat transfers instantly to the soil mass and neglects any heat transfer to the plants or other objects in the greenhouse.

3.2 Discharging reservoir heat to the ambient air

Ninety percent of the energy required to operate the greenhouse is used to discharge the accumulated reservoir heat to the ambient air during the night and morning. The simulation discharges reservoir heat between 19:00 in the evening and 9:00 in the morning. During this period the sides of the tunnel are closed to the remainder of the greenhouse volume and the ports to the heat exchanger on either end of the tunnel are open. The tunnel and heat exchanger form an isolated system that is simulated independently from the remainder of the greenhouse. **Figure 6** shows the model of ambient temperature T_o used by the simulation. The most critical feature for discharging heat is the minimum diurnal temperature, assumed to be 11.8°C in this example, at 7:00.

If the ambient temperature is at least 2°C less than the reservoir temperature, pumps are activated to circulate 450 L s^{-1} into the droplet dispensers and fans are

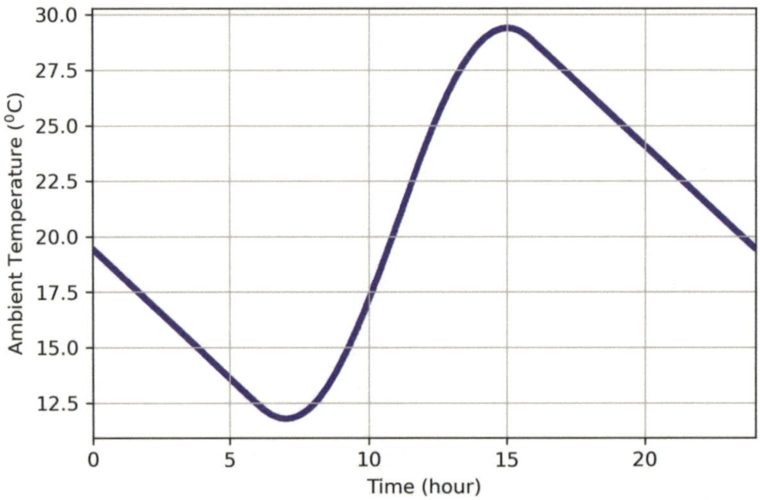

Figure 6.
Model of the ambient temperature for the simulation over 24 hours. The temperature reaches a minimum of 11.8° C at 7:00.

activated to pull air through the heat exchanger tubes and to push the cool, saturated air through the falling droplets down the length of the tunnel. The fan power is adjusted so that the airspeed in the heat exchanger tubes v_x is linearly proportional to the difference between the reservoir water temperature T_r and the ambient air temperature T_o, reaching a maximum of 5 m s^{-1} during the morning. The airspeed v_T in the tunnel is a factor 2.20 less than v_x and determined by the ratio of cross-sectional areas of the tunnel and heat exchanger tubes.

The air circulating through the tunnel and heat exchanger is always saturated. As cool air moves down the tunnel the warm reservoir droplets transfer both heat and water vapor to the air at the surface of the droplet, but as the water vapor diffuses away from the warm surface of the droplet it is assumed to recondense as fog, so that both the sensible and latent heat contributed from the droplet raise the temperature of the tunnel air. When the warmed air at the end of the tunnel enters the heat exchanger, it loses heat to the ambient air through the wall of the heat exchanger tube, and water condenses within the tube as the temperature and saturation vapor pressure fall. The simulation follows a Lagrangian air parcel through the tunnel to calculate the heat transfer from the reservoir water to the air, and a second Lagrangian parcel through a heat exchanger tube to calculate the transfer of heat from the air in the heat exchanger tube to the outside.

The tunnel Lagrangian parcel is a lamina of tunnel cross section with volume $\Delta V_T = h\Delta W\Delta Z$, where h = 1.54 m is the height between the reservoir water surface and the bottom of the droplet dispenser trays, ΔZ = 10 m is the width of the tunnel, and ΔW = 5 cm is the thickness of the lamina along the direction of tunnel airflow. During the small time step $dt = \Delta W/v_T << \Delta t$, the heat $dQ_{T,\mathrm{Cnv}}$ that is transferred convectively from the falling droplets to the tunnel air is calculated within the parcel volume ΔV_T. The time step dt that is used to update the Lagrangian parcel is much smaller than the simulation time step Δt that is used to update Eqs. (1)–(3). The reservoir droplets also release the water vapor mass $-dM_r$ into the parcel air. The temperature increase $dT_{T,a}$ of the parcel air due to both sensible and latent heat is then:

$$dT_{T,a} = \left(dQ_{T,\mathrm{Cnv}} - H_v dM_r\right)/\left(\Delta V_T \rho_a C_{pa}\right) \tag{4}$$

After the parcel has traversed the entire length of the tunnel it has accumulated the heat δQ_T so that heat is removed from the reservoir at the rate $\delta Q_T/dt$. Therefore, during the simulation time step Δt the reservoir temperature changes by:

$$\Delta T_r = -\left(\frac{\Delta t}{dt}\right)\frac{\delta Q_T}{A_r d_r \rho_w C_{pw}} \tag{5}$$

Within the heat exchanger tube the simulation follows a cylindrical Lagrangian parcel with volume $\Delta V_x = \pi D_0{}^2 \Delta L/4$ where $\Delta L = 3$ cm is the length of the parcel and D_0 is the inner diameter of the tube. The time step for the heat exchanger tube simulation is $dt = \Delta L/v_x < < \Delta t$.

The heat exchanger tube enables the forced convection of heat to the ambient air through the total heat transfer coefficient h_c. There are three contributions to h_c that each represent a resistance to heat transfer out of the tube [22]:

$$\frac{1}{D_0 h_c} = \frac{1}{D_0 h_0} + \frac{\log(D_1/D_0)}{2k_{01}} + \frac{1}{D_1 h_1} \tag{6}$$

The first term on the right hand side with heat transfer coefficient h_0 computes the convection across the boundary layer of air flowing within the tube. The third term with heat transfer coefficient h_1 computes the convection across the boundary layer of the outside crossflow air stream flowing through the heat exchanger tube bank. Outside fans pull air across the tube bank at the same speed $v_1 = v_T$ as the airspeed through the tunnel.

The second term with thermal conductivity coefficient k_{01} computes the heat conduction through the wall of the tube. For PETG tubing, $k_{01} = 0.0029\text{W cm}^{-1}\text{ K}^{-1}$. Although the thermal conductivity of plastic tubing is very low, metal tubing for the heat exchanger was rejected as impractical because of cost. Because the 0.5 mm tube wall is very thin, the thermal conductivity through the wall contributes only about 3% of the total resistance to heat transfer in Eq. (6).

The heat transfer coefficients are calculated from dimensionless Nusselt numbers: $h_0 = \mathrm{Nu}_0 k_a/D_0$ and $h_1 = \mathrm{Nu}_1 k_a/D_1$, where k_a is the thermal conductivity for air. For turbulent air flow [23, 24]:

$$\begin{aligned} \mathrm{Nu}_0 &= 0.023\,\mathrm{Re}_0^{0.8}\mathrm{Pr}^{0.33} \\ \mathrm{Nu}_1 &= 0.33\,\mathrm{Re}_1^{0.6}\mathrm{Pr}^{1/3} \end{aligned} \tag{7}$$

$\mathrm{Re}_0 = D_0 v_x \rho_a/\mu_a$ is the Reynolds number for air flow inside the tube, where μ_a is the viscosity of air, and $\mathrm{Re}_1 = D_1 v_1 \rho_a/\mu_a$ is the Reynolds number for the outside cross flow of air within the tube bank. The Prandtl number is given by $\mathrm{Pr} = C_{pa}\mu_a/k_a$. Note that the temperature loss down the tube is only a weak function of the airspeed v_x: although the residence time for a parcel of air within a tube section of length ΔL is $dt = \Delta L/v_x$, the heat transfer coefficient is proportional to $v_x^{0.8}$. Hence the total rate of heat transfer roughly scales with v_x so that high speed turbulent flow through the heat exchanger tubes is preferred.

The temperature drop ΔT within a tube parcel during a time step dt depends upon the heat dQ_w transferred from the air through the wall of the tube and on the heat of fusion dQ_m released into the volume of the parcel as water vapor in the saturated air condenses due to the drop in temperature:

$$\rho_a C_{pa} \Delta V_x \Delta T = dQ_w - dQ_m \tag{8}$$

In Eq. (8), the temperature drop ΔT, dQ_w, and dQ_m are all unsigned positive quantities. dQ_w and dQ_m are given by:

$$\begin{aligned} dQ_w &= h_c(\pi D_0 \Delta L)(T - T_o)dt \\ dQ_m &= \left(\pi D_0^2 \Delta L/4\right)\rho_a H_v \Delta T \frac{dX}{dT} \end{aligned} \tag{9}$$

where X and T are the mixing ratio and temperature of the tube parcel air. Eqs. (8) and (9) may be combined to solve for the temperature drop ΔT during the time interval dt:

$$\Delta T = \frac{4h_c(T - T_o)dt}{\rho_a D_0 \left(C_{pa} + H_v \frac{dX}{dT}\right)} \tag{10}$$

The derivative dX/dT is obtained by differentiating the expression for the mixing ratio in terms of the saturation vapor pressure $P_s(T)$:

$$X(T) = \frac{M_w P_s(T)}{M_a(P_{\text{atm}} - P_s(T))} \tag{11}$$

where M_a and M_w are the gram molecular weights of air and water and P_{atm} is one atmosphere of pressure [21].

3.3 Greenhouse temperature, humidity, and VPD over a 24 hour cycle

Figure 7 shows the temperature of the greenhouse air and reservoir over a 24-hour cycle. The sun sets at 19:00 and the reservoir water is warm from the previous day. The sides of the tunnel are closed off from the remainder of the greenhouse volume and the tunnel is opened at each end to the ducts that lead to the heat exchanger tube bank in preparation for discharging the accumulated reservoir heat. The ambient temperature decreases (**Figure 6**) and the temperature of the greenhouse air also decreases as heat is conducted through the greenhouse cover. At 22:47 the ambient air temperature has fallen to 2°C less than the reservoir temperature and the fans and dispensers are activated in the tunnel to discharge the reservoir heat. The temperature in the reservoir continues to drop until 9:00 when the heat discharge period ends. During this time the sides of the tunnel are opened to the remainder of the greenhouse volume and the ports to the heat exchanger at the ends of the tunnel are closed. Droplets are dispensed to cool the greenhouse air and the reservoir temperature begins to rise.

The sun rises and begins to heat the greenhouse at 7:00. The temperature begins to climb rapidly until the daytime droplet activation period begins at 9:00. After 9:00 the temperature continues to climb more slowly as the insolation increases and the reservoir temperature increases, reducing the efficacy of the droplet system. At 15:30 the air temperature begins to gradually subside as the solar insolation drops. At 17:00 the droplet flow is reduced by 60% from 100 L s^{-1} to 40 L s^{-1} and there is a cusp in the air

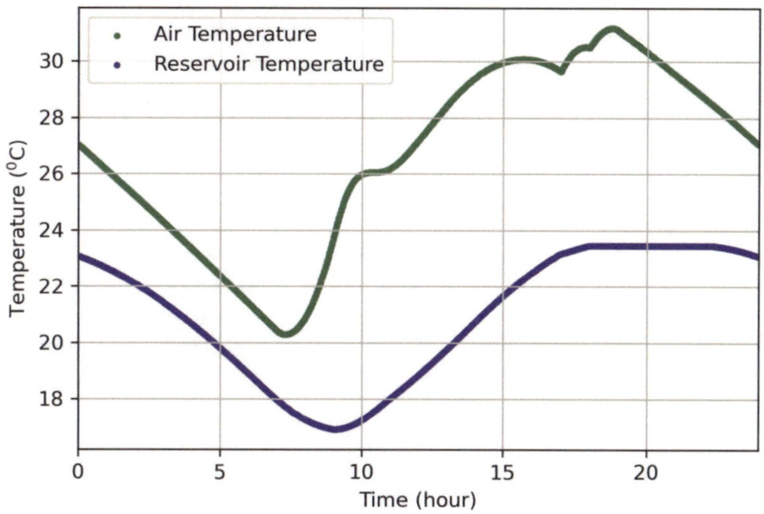

Figure 7.
Greenhouse air and reservoir water temperatures over a 24-hour cycle.

temperature curve when the temperature starts to rise. At 18:00 the droplets are turned off entirely creating a second cusp in the curve.

Figure 8 is a plot of the relative humidity (RH) of the greenhouse air. During the night the RH gradually drops because the drop in air temperature is more than compensated with the removal of water vapor by the condensing tubes on the roof of the greenhouse. At 7:00 light begins to enter the greenhouse and the plants transpire, overwhelming the condensing tubes and causing the RH to rise rapidly. The humidity drops sharply at 9:00 when the droplet dispensers are activated for daytime cooling. The rising greenhouse temperature and falling efficacy of the droplet cooling compensate one another to keep the RH roughly constant until 15:00, when the RH begins to rise. At 17:00 the droplet flow is reduced by 60%, causing a small jump in RH that is

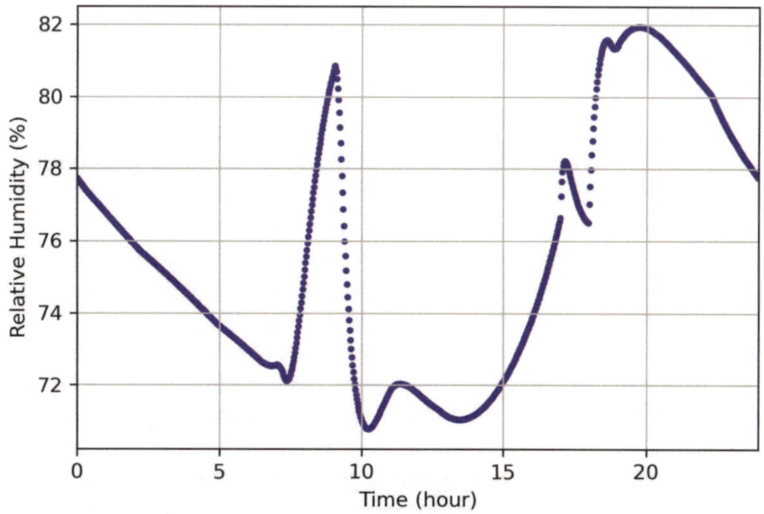

Figure 8.
Greenhouse relative humidity over a 24-hour cycle.

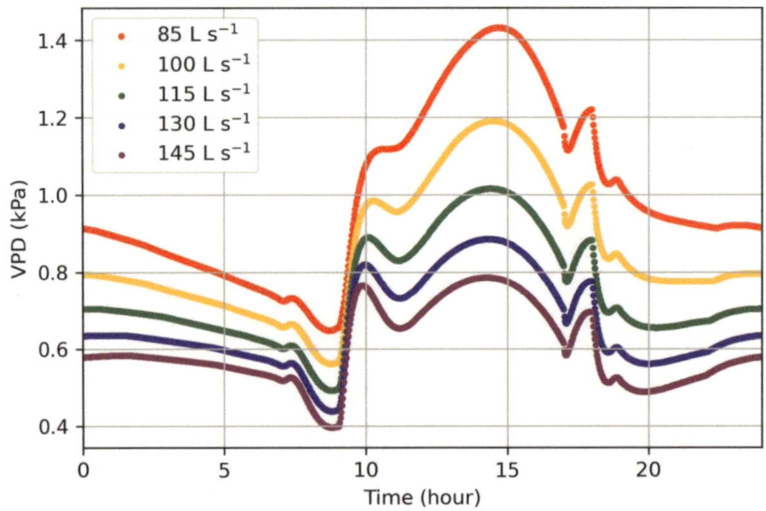

Figure 9.
Sensitivity of the vapor pressure deficit to the daytime droplet flow in the greenhouse over a 24-hour cycle. The previous plots were created for a daytime droplet flow of 100 L s^{-1}.

quickly counteracted by the rise in temperature. At 18:00 when the droplets are shut off the RH rises rapidly, but this is counteracted by the steep drop in plant transpiration as the sunlight disappears, so that the condensing tubes begin reducing the humidity after 20:00.

The greenhouse temperature and relative humidity may be combined to compute VPD. The VPD controls the transpiration of the plants, and it has been argued that the VPD is the parameter most relevant to the comfort of the plants [25, 26]. **Figure 9** is a plot of the greenhouse VPD for several values of the daytime droplet flow, demonstrating that the VPD may be tuned for this greenhouse design by adjusting the daytime droplet flow. VPD values in the range of 0.4–1.3 kPa are optimal for greenhouse cultivation [25].

4. Discussion and conclusions

The simulation predicts that 191.5 kWh are required to operate the fans and pumps during a 24-hour cycle to cool a 1000 m^2 greenhouse, assuming the conditions specified in **Figures 3** and **6** for the incoming solar radiation and the ambient temperature. 62.4% of this energy is used to operate the pumps. Once the reservoir is cold, the greenhouse air may be cooled during the day with a smaller expenditure of energy. 90.3% of the energy required to operate the greenhouse is used to discharge the accumulated reservoir heat using the air-to-air heat exchanger, or 173 kWh. The strategy to leverage the ambient temperature difference between the morning and afternoon significantly reduces the energy cost compared to using an air-cooled chiller, which requires 6.9 times the energy for a coefficient of performance of 4 [21]. The maintenance of the proposed cooling system would be simpler than the maintenance of a chiller, requiring the occasional replacement of pumps, fans, heat exchanger tubes, and tiles for the dispensers.

The energy cost per cultivated area for the closed greenhouse is far greater than using a conventional ventilated greenhouse cooled by a fan-pad system. Cooling a

1000 m^2 greenhouse evaporatively with a fan-pad system for 8 h during the day would require 14.8 kWh for the fans, assuming an airflow of 60 $m^3\ s^{-1}$ [21], a factor 13 less than the closed greenhouse design proposed here. Furthermore the conventional greenhouse does not require the costs for constructing the closed greenhouse cooling system including pumps, extra fans, reservoirs, dispensers, and heat exchanger tubes. The significant additional costs of constructing and operating a closed greenhouse enable conserving fresh water, maintaining an arbitrarily high concentration of CO_2 during the afternoon period of peak photosynthesis, and sequestering captured CO_2 into biomass.

Civilization is entering an era of abundant renewable energy but diminishing freshwater resources. Photovoltaics and battery storage have enormous potential for future innovation, and renewable energy costs will continue to fall from economy of scale as solar and wind replace fossil fuels. However freshwater resources are increasingly precious. The conventional fan-pad greenhouse used in the comparison above consumes 7.9 $m^3\ d^{-1}$ of water for evaporative cooling assuming an ambient temperature of 33°C and ambient relative humidity of 37% [21]. For a limited number of greenhouses this amount of water may not be a concern, but for many square kilometers under greenhouse cultivation in the desert a water loss this high becomes impractical.

It may be possible to reduce the cost of cooling system components by using recycled plastic. The reservoirs, heat exchanger tubes, and dispensers are made from thermoplastics that may be melted and reformed repeatedly. Recycling used plastic products into new products is the most desired means of disposal [27], so that extensive closed greenhouse construction might become a useful output stream for plastic waste.

The enhancement of photosynthesis at higher temperatures in the presence of elevated levels of CO_2 is fortuitous for closed greenhouses because of the energy demand by the cooling system to reduce the temperature and dehumidify the air. If the greenhouse temperature is allowed to increase, then the greenhouse relative humidity must also increase to maintain the same VPD for plant transpiration.

The photosynthesis enhancement at elevated temperature and CO_2 has been investigated for tomatoes [11, 12] and other plants [10, 13, 14]. The biochemical mechanism for this enhancement, and the mechanism for inducing damage to photosynthesis under significant heat stress when elevated levels of CO_2 are not present, remain active areas of investigation [12]. At the beginning of the Calvin cycle in photosynthesis, CO_2 attaches to the sugar substrate RuBP (ribulose bisphosphate), catalyzed by the enzyme rubisco (ribulose bisphosphate carboxylase). As the temperature increases a competing reaction becomes more favorable, where oxygen attaches to RuBP instead of CO_2, leading to photorespiration rather than photosynthesis. By increasing the concentration of CO_2 the photorespiration reaction may be suppressed so that photosynthesis can take advantage of the increased activity of rubisco at higher temperature. This mechanism holds at all light intensities, although the effect is enhanced with increasing light intensity [11].

If the temperature increases too much it may damage plant photosynthesis if the CO_2 concentration is not also increased. One hypothesis [12] argues that increased heat stress reduces the stomatal conductance, reducing the CO_2 available to the Calvin cycle and effectively reducing or interrupting electron transport to rubisco and RuBP, reducing both the activity of rubisco and the regeneration of RuBP. Thylakoid electron transport away from the PSII reaction center (PSII RC) is effectively blocked or reduced, causing excessive reduction of the acceptor which damages the PSII RC.

Elevating ambient CO_2 restores the flow of electrons from the PSII RC to the Calvin cycle, restoring the redox balance along the electron transport path so that damage to the PSII RC is prevented.

Closed greenhouses could extend agriculture in higher elevation deserts into regions far beyond what might be currently irrigated near a river or lake. In addition to enhancing yields for standard greenhouse crops such as tomatoes and peppers, closed greenhouses could cultivate woody plants to sequester CO_2 into lumber and biochar. Bamboo for example is fast-growing and very responsive to enhanced concentrations of CO_2 [28]. A large-scale program of CO_2 sequestration into biomass might best be implemented in selected climates that are extremely favorable for deployment of the cooling system, such as the Altiplano plateau in Bolivia, Peru, and Chile.

The cooling system functions equally well in high humidity climates where evaporative cooling may be impractical. Coffee may be a suitable closed greenhouse crop because it prefers high humidity and shade, and a greenhouse would help protect the plants from unfavorable climate change induced conditions such as droughts, heat waves, and pests. Enhanced CO_2 concentrations also allow coffee plants and bean quality to endure supra-optimal temperatures during the day and night [13]. The coffee plant shows both reduced photorespiration in the presence of elevated CO2 and increased thylakoid electron transport [14].

Future research will include building and testing a full prototype and further simulations to explore variations and improvements to the design. One variation under consideration is an aquaponic system, where cold water fish are raised in tall cisterns that serve as reservoirs, and nutrient rich reservoir water is circulated between the reservoirs and hydroponically grown vegetables or a deep water culture of lettuce or rice. Another variation is a closed greenhouse for northern climates, where the heat exchanger tubes are optionally enclosed within the greenhouse volume during the winter. Heat captured by the reservoir during the day is released by the heat exchanger into the cold greenhouse air at night to prevent the plants from freezing.

In summary, the simulation has demonstrated that it is possible to cool a closed greenhouse with a large reservoir of water and an air-to-air heat exchanger comprised of thin-walled plastic tubes. The reservoir volume used in the design is $A/2\ m^3$, where A is the area of the cultivated region in m^2. The closed greenhouse requires about an order of magnitude more energy to cool than a conventional greenhouse cooled evaporatively using a fan-pad system. Hence the cooling system relies upon a high expenditure of electric power to conserve water and to maintain high concentrations of CO_2 during the day to enhance yields. The summer insolation and minimum daily temperature are the most important parameters that decide the size and cost of the cooling system, so that regions where the greenhouse design may be deployed are restricted to climates with cool morning temperatures. The VPD within the greenhouse may be tuned by adjusting the droplet flow that cools the greenhouse during the day. Deployment of closed greenhouses in regions with very low minimum diurnal temperatures ($< 8°C$) during summer potentially allow profitably sequestering CO_2 into biomass.

Acknowledgements

The author gratefully acknowledges the advice and encouragement of Professor J. Heiner Lieth at the University of California, Davis, and his consent to conduct heat and mass transfer experiments using a greenhouse on the Davis campus.

Conflict of interest

The author declares no conflict of interest.

Author details

Geordie Zapalac
Independent Researcher, Santa Clara, CA, USA

*Address all correspondence to: geordie.zapalac@gmx.com

References

[1] Mariem SB et al. Climate change, crop yields, and grain quality of C_3 cereals: A meta-analysis of [CO_2], temperature, and drought effects. Plants. 2021; **10**(6):1052-1070. DOI: 10.3390/plants10061052

[2] Capon SJ, Stewart-Koster B, Bunn SE. Future of freshwater ecosystems in a 1.5°C warmer world. Frontiers in Environmental Science. 2021;**9**:784642. DOI: 10.3389/fenvs.2021.784642

[3] Burrell AL, Evans JP, De Kauwe MG. Anthropogenic climate change has driven over 5 million km^2 of drylands towards desertification. Nature Communications. 2020;**11**:3853. DOI: 10.1038/s41467-020-17710-7

[4] Lowery B et al. The 2008 Midwest flooding impact on soil erosion and water quality: Implications for soil erosion control practices. Journal of Soil and Water Conservation. 2009;**64**(6): 166A. DOI: 10.2489/jswc.64.6.166A

[5] Eslami S et al. Projections of salt intrusion in a mega-delta under climatic and anthropogenic stressors. Communications Earth & Environment. 2021;**8**:142. DOI: 10.1038/s43247-021-00208-5

[6] Climate Change 2022: Mitigation of Climate Change. Working Group III Contribution to the IPCC Sixth Assessment Report. Geneva, Switzerland: Intergovernmental Panel on Climate Change; 2023

[7] Ehlig-Economides C, Economides MJ. Sequestering carbon dioxide in a closed underground volume. Journal of Petroleum Science and Engineering. 2010;**70**:123-130. DOI: 10.1016/j.petrol.2009.11.002

[8] Akinnikawe O, Ehlig-Economides CA. Geologic model and fluid flow simulation of Woodbine aquifer CO_2 sequestration. International Journal of Greenhouse Gas Control. 2016;**49**:1-13. DOI: 10.1016/j.ijggc.2016.02.014

[9] IEA. Storing CO2 Through Enhanced Oil Recovery. Paris: IEA; 2015. Available from: https://www.iea.org/reports/storing-co2-through-enhanced-oil-recovery [Accessed: August 30, 2023], License:CC BY 4.0

[10] Jurik TW, Weber JA, Gates DM. Short-term effects of CO_2 on gas exchange of leaves of Bigtooth Aspen (*Populus grandidentata*) in the field. Plant Physiology. 1984;**75**:1022-1026. DOI: 10.1104/pp.75.4.1022

[11] Nilsen S, Hovland K, Dons C, Sletten SP. Effect of CO_2 enrichment on photosynthesis, growth and yield of tomato. Scientia Horticulturae. 1983;**20**: 1-14. DOI: 10.1016/0304-4238(83)90106-1

[12] Pan C, Ahammed GJ, Li X, Shi K. Elevated CO improves photosynthesis under high temperature by attenuating the functional limitations to energy fluxes, electron transport and redox homeostasis in tomato leaves. Frontiers in Plant Science. 2018;**9**:1-11. DOI: 10.3389/fpls.2018.01739

[13] DaMatta FM, Rahn E, Läderach P, Ghini R, Ramalho JC. Why could the coffee plant crop endure climate change and global warming to a greater extent than previously estimated? Climate Change. 2019;**152**:167-178. DOI: 10.1007/s10584-018-2346-4

[14] Ramalho JC et al. Sustained photosynthetic performance of *Coffea*

spp. under long-term enhanced [CO_2]. PLoS One. 2013;**8**(12):e82712. DOI: 10.1371/journal.pone.0082712

[15] Vadiee A, Martin V. Energy analysis and thermoeconomic assessment of the closed greenhouse–The largest commercial solar building. Applied Energy. 2012;**102**:1256-1266. DOI: 10.1016/j.apenergy.2012.06.051

[16] Opdam JJG, Schoonderbeek GG, Heller EMB, de Gelder A. Closed greenhouse: A starting point for sustainable entrepreneurship in horticulture. Acta Horticulturae. 2005; **691**:517-524. DOI: 10.17660/ ActaHortic.2005.691.61

[17] Zaragoza G, Buchholz M. Closed greenhouses for semi-arid climates: Critical discussion following the Watergy prototype. Acta Horticulturae. 2008;**797**:37-42. DOI: 10.17660/ ActaHortic.2008.797.2

[18] Buchholz M, Buchholz R, Jochum P, Zaragoza G, Para JP. Temperature and humidity control in the Watergy greenhouse. Acta Horticulturae. 2006; **719**:401-408. DOI: 10.17660/ ActaHortic.2006.719.45

[19] Novarbo. Heat Reuse™ [Internet]. Available from: https://www.novarbo.fi/en/greenhouse-technology/heat-reuse.html [Accessed: August 30, 2023]

[20] Huttunen J. Closed greenhouse cooling with water droplet curtain. Acta Horticulturae. 2011;**893**: 1043-1047. DOI: 10.17660/ ActaHortic.2011.893.118

[21] Zapalac G. Simulation of a convectively-cooled unventilated greenhouse. Computers and Electronics in Agriculture. 2022;**193**:106563. DOI: 10.1016/j.compag.2021.106563

[22] Bird RB, Stewart WE, Lightfoot EN. Transport Phenomena. New York: Wiley; 1960. p. 780. DOI: 10.1002/ aic.690070245

[23] Edward DK, Denny WK, Mills AF. Transfer Processes: An Introduction to Diffusion, Convection, and Radiation. New York: McGraw-Hill; 1979. p. 421. ISBN: 13: 9780070190412

[24] Khan WA, Culham JR, Yovanovich MM. Convection heat transfer from tube banks in crossflow: Analytical approach. International Journal of Heat and Mass Transfer. 2006;**49**:4831-4838. DOI: 10.1016/j.ijheatmasstransfer. 2006.05.042

[25] Shamshiri RR et al. Review of optimum temperature, humidity, and vapour pressure deficit for microclimate evaluation and control in greenhouse cultivation of tomato: A review. International Agrophysics. 2018;**32**: 287-302. DOI: 10.1515/intag-2017-0005

[26] Zolnier S et al. Psychometric and ventilation constraints for vapor pressure deficit control. Computers and Electronics in Agriculture. 2000;**26**: 343-359. DOI: 10.1016/S0168-1699(00) 00084-3

[27] Evode N et al. Plastic waste and its management strategies for environmental sustainability. Case Studies in Chemical and Environmental Engineering. 2021;**4**:100142. DOI: 10.1016/j.cscee.2021.100142

[28] Grombone-Guarantini MT et al. Atmospheric CO_2 enrichment markedly increases photosynthesis and growth in a woody tropical bamboo from the Brazilian Atlantic Forest. New Zealand Journal of Botany. 2013;**51**:275-285. DOI: 10.1080/0028825X.2013.829502

Section 3

Assessing Environmental Impact and Mitigation Strategies

Chapter 7

Life Cycle Assessment of Natural Gas Power Plant: Calculation of Impact Potentials

Oludolapo Akanni Olanrewaju, Oluwafemi Ezekiel Ige, Busola Dorcas Akintayo and Ahad Ali

Abstract

Natural gas is a growing energy source worldwide, with its market share increasing steadily. It is one of the primary fuels used in electricity production. Its high thermodynamic quality and low environmental impact make it the fastest growing energy source in the global energy sector. Natural gas is a relatively clean and efficient fuel, making it a good choice for electricity production and heating. Using natural gas in gas power plants and industrial thermal applications will reduce harmful pollutants. Despite its significance, it is crucial to understand its potential impact on the electricity supply. The objective of this study is to conduct a life cycle assessment from cradle-to-gate of a natural gas power plant to understand the impact on the global warming (GWP) potential, freshwater eutrophication potential (FEP) and terrestrial acidification potential (TAP) categories when producing 1 kWh of electricity. Using the SimaPro (version 9.2) software package and Rest of the World data to model the cradle-to-gate scenario, the study found that the processing of natural gas is the most crucial stage in all three impact categories, making it the hotspot (37-95%) for GWP, FEP and TAP, with CO_2 contributing the most at the GWP, PO_4 at FEP and NOx at TAP.

Keywords: natural gas plant, life cycle assessment, global warming, eutrophication, acidification

1. Introduction

Generally, global electricity generation comes from fossil fuels, that is, coal, oil, and natural gas, with thermal power plants producing most of this electricity [1, 2]. The power generation sector has been identified as the highest contributor to global greenhouse gas (GHG) emissions [3]. This trend is likely to continue as the demand for electricity increases. Global electricity generation modes are changing substantially due to depleting fossil resources and a looming climate emergency, putting pressure on countries to adopt low-carbon energy policies. According to the

Intergovernmental Panel on Climate Change (IPCC), global economy electrification and grid rapid carbon footprint are potential measures for reducing greenhouse gas emissions (GHG) and keeping global warming at 1.5°C or below 2°C [4]. In the global energy sector, energy activities include extraction, conversion, intermediate and final energy use, accounting for approximately 75% of GHG emissions [5, 6], primarily due to burning coal, natural gas, and oil. Most of these burnings are used to generate electricity today. Replacing coal and oil with natural gas in power generation and industrial thermal applications will help reduce harmful pollutants. This emissions reduction qualifies natural gas as a cleaner fuel. With concerns about air quality and climate change, natural gas has a future role in global energy supply as a cleaner fossil fuel that is still abundant in supply, aided by the fact that renewable energy options remain limited in ability to scale up while cost-effective zero-carbon options can be harder to find in some [7, 8].

Power plants powered by natural gas are cheap and easy to build. In addition, they have exceptionally high thermodynamic efficiency compared to other conventional power plants. Natural gas power plants are less polluting than coal and oil plants because they emit less nitrogen oxides (NO_x), sulfur oxide (SO_x) and particulate matter (PM) [9]. Natural gas is used as their fuel in natural gas power plants to produce electricity. Natural gas is composed of hydrocarbons that exist naturally and is frequently discovered in petroleum and coal deposits, as well as in the form of hydrates on the ocean floor. The gas is formed by the decomposition of organic matter [10]. The gas can be liquefied by cooling it to a temperature of −162°C at atmospheric pressure [11]. The main constituent of natural gas is methane (CH_4) mixed with smaller amounts of, moisture or water vapor, nitrogen propane, ethane, carbon oxide (CO_2), helium, and hydrogen sulfide [12]. The thermodynamic properties of natural gas from its combustion produce negligible amounts of sulfur, mercury, particulates and small quantities of NOx, making natural gas a cleaner fuel than gasoline and diesel [7]. A natural gas engine refers to a mechanical engine that utilizes natural gas as its primary fuel source for the generation of power, whether in the form of mechanical energy or electrical energy.

Natural gas has several advantages over other fossil fuels, including affordability, accessibility, environmental friendliness, compatibility with conventional spark and compression ignition engines, and low operating costs [13]. These advantages make natural gas appealing for power generation and other engine applications. Similarly, natural gas power plants offer various advantages, such as effective combustion, cost-effectiveness, adherence to environmental standards, enough availability and supply, and the generation of cleaner energy [13]. Despite these advantages, examining the environmental impact is crucial before deciding whether to invest in such a resource. Some of those studies include energy transition and air pollution impact [14], biodiesel production impact [15] and impact of electricity options, which includes renewables [16–19]. All electricity production has an environmental impact throughout its entire life cycle, from the production stage to end-use.

Life cycle assessment (LCA) methodology has been widely used to evaluate the environmental impacts of energy systems and has gained popularity in areas where environmental impact is of concern [20–22]. The primary objective of this study is to assess the environmental ramifications associated with the operation of a gas power plant through the utilization of a Life Cycle Assessment (LCA) methodology. There have been several LCIA studies conducted in the past, including ReCiPe, CML, TRACI, IMPACT2002+ and IPCC. Several impact categories ranging from one up to

10 environmental categories have been investigated [23, 24]. Also, the impact categories seem to differ across studies. Impact categories such as global warming potentials (GWP), ozone depletion potentials (ODPs), human toxicity potentials (HTP), acidification potentials (AP) and eutrophication potentials (EPs) are more often used at the midpoint characterization in LCA studies for electricity generation [25]. Ecoinvent, a database that provides information on potential environmental impacts, was the most widely used database across the research. Several LCA studies have been conducted to assess the different electricity generation in specific countries such as Poland [26], Portugal [17, 27], United Kingdom (UK), Belgium [28, 29], Denmark [30, 31], Pakistan [32], China [14] and Brazil [33, 34].

Wu et al. [14] used LCA in the integration model to quantitatively evaluate its environmental impact under three policy scenarios from 2016 to 2050. Under the deep-level cut of CO_2 emission to achieve emission reduction and carbon neutrality in China from energy transition and air pollution. CO_2, $PM_{2.5}$, PM_{10}, CO, NO_X and SO_2 were deemed to reduce by more than 71.4% compared to the 2016 records. There would also be an 81.8% - 88.5% decrease in the global warming potential, human toxicity potential, petrochemical ozone creation potential, particulate formation potential and acidification potential. Alizadeh and Avami [15] compared biodiesel production from palm oil and multi-feedstock using LCA in Indonesia to understand the environmental impact of biodiesel production. There was a higher environmental impact from using multi-feedstock at the plantation stage, with 9.89tCO_2 GHG emission per tonne in the land use of scrubbed plantation to just -3.42 tCO_2 of palm oil plantation on biodiesel production.

Roinioti and Koroneos [16], looked at the sustainability aspect of the electricity options (Lignite, Combined Cycle Gas Turbine – CCGT, Hydro, Small hydro, Wind, PV and Biogas-biomass) in Greece through LCA considering the environmental, economic, and social dimensions. This study only showed interest in the environmental viewpoint. Looking at the seven sustainable indicators under environment, Biomass (54.26 g CO_2 equ./Kw) had the worst impact under the Global Warming Potential indicator among the renewables, while Lignite station (1067.15 g CO_2 equ./kWh) had the overall worst impact followed by CCGT station (509.96 gCO_2 equ./kWh). Lignite exhibited the worst impact for Acidification Potential (2989.89 mg SO_2 equ./kWh) and for Tropospheric Ozone Precursor Potential (2644.63 mg TOPP equ./kWh). Biomass-biogas exhibited the worst impact for both Eutrophication Potential (891.61 mg PO_4 equ./kWh) and Photochemical Oxidation Potential (0.49 mg ethene equ./kWh). On the Ozone Depletion Potential, CCGT exhibited the worst with a 99.98 μg CFC-11equ./kWh value.

Kabayo et al. [17], performed a life cycle sustainability assessment on Portugal's electricity (coal, natural gas, large hydro and small hydro, wind, and photovoltaic) generation from the environmental and socioeconomic impacts. Interest remained in the environmental impact for the essence of this study. The highest impact of the sources of electricity is in brackets next to the sustainable indicators considered: metal depletion (wind), fossil fuel depletion (coal), global warming (coal), ozone depletion (natural gas), terrestrial acidification (coal), freshwater eutrophication (coal), aquatic acidification (coal), water scarcity footprint (large hydro), and toxicity (photovoltaic). The environmental impact of renewable energy generation systems was analyzed comparatively using life cycle assessment for Europe, North America, and Oceania [18]. The energy-generation technologies included wind, photovoltaic, biomass, and hydropower. The indicators for the impact were ozone layer depletion, freshwater aquatic ecotoxicity and marine aquatic ecotoxicity, abiotic depletion, global warming, photochemical oxidation, acidification and eutrophication. Of all the generation systems, biomass was impacted the most, followed by photovoltaic.

Mahmud et al. [19] conducted a comparative LCA of solar-photovoltaic and solar-thermal systems to determine their environmental impact using 16 indicators. The result indicated that the solar-thermal system is more impactful compared to the photovoltaic. The study recommended careful component selection and reducing the impact related to solar panels, batteries and heat storage for better environmental performance. This study aimed to identify the hotspots of environmental impacts associated with natural gas power plants. The authors have solely considered and concentrated on Global Warming, Eutrophication and Acidification impact categories due to climate change using SimaPro (version 9.2) software package to assess the environmental impact of producing 1KWh of electricity from natural gas on human health. This study will cover the potential impact of producing 1 kWh of electricity from natural gas in detail on views from global warming, acidification and eutrophication.

2. Materials and methods

2.1 Environmental impact through life cycle assessment (LCA)

LCA evaluates the environmental impact emanating from a production process for a product stipulated in ISO 14040 within the process categories and boundaries [35]. The implementation of the aforementioned concept exhibits variability across different studies, contingent upon the explicitly stated objectives of each respective investigation. The LCA's operation occurs at the following stages (**Figure 1**), which are (1) the goal and scope definition, (2) life cycle inventory (LCI), (3) life cycle impact assessment (LCIA) and (4) results and interpretation [36] through the energy and the materials used been quantified and the release of the wastes to the environment [37]. LCA has been used to investigate the environmental impact of the natural gas power plant. The modeling of the impacts was done by using SimaPro (version 9.2) software package.

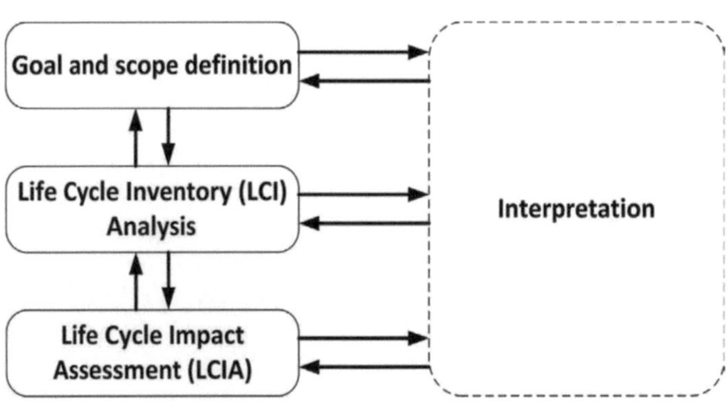

Figure 1.
ISO 14040 LCA framework [36].

2.2 Definition of goal and scope

The first part of a standardized LCA is goal and scope definition, which includes a description of essential choices, such as why the LCA is being done, a comprehensive elucidation of the product or process along with its life cycle, accompanied by a depiction of the system boundaries. It is possible to comprehend why LCA is undertaken by stating the study's purpose. However, various factors influence the system boundary, including the study's aim, assumptions and intended audience. ISO 14040 recommends that the circumstances used to define the system boundary be defined and justified in the study's scope [36]. The goal of this study is to investigate, quantify and compare the effect on the environment of producing 1kWh of high-voltage electricity from natural gas in a power plant. This is a 'cradle-to-gate' approach [37, 38]. **Figure 2** shows the boundaries for the system.

The boundaries include the extraction phase (conventional onshore source) to the electricity production in a power plant's busbar. This study did not consider inventory data on distribution, use, disposal/end-of-life, or waste treatment. In addition, the midpoint LCIA technique is used in this study. The functional unit of this work is 1 kWh net electricity produced from natural gas. This study normalized all the inputs to the functional unit.

The study focuses on the three impact categories from the 18 impact categories presented in the midpoint analysis characterization result. These are the Global Warming potential (GWP), Freshwater Eutrophication Potential (FEP) and Terrestrial Acidification Potential (TAP). At the time of the study, primary data was inevitably unavailable; hence, secondary data from the Ecoinvent database 3.7.1 [39, 40] was employed. Assumptions and uncertainties have been adequately adjusted as specified by the dataset documentation and data has been interpolated accordingly.

2.3 Life cycle inventory (LCI)

The Life Cycle Inventory (LCI) is the data collection phase of the LCA since analysis entails compiling input and output inventory data that are consistent with the product under consideration and cover various environmental factors [41]. The comprehensive data encompasses all inputs and outputs of the system, encompassing materials, resources, energy, and emissions across the whole life cycle of the process or product [42]. The data for this study was derived from the Ecoinvent (V 3.7.1)

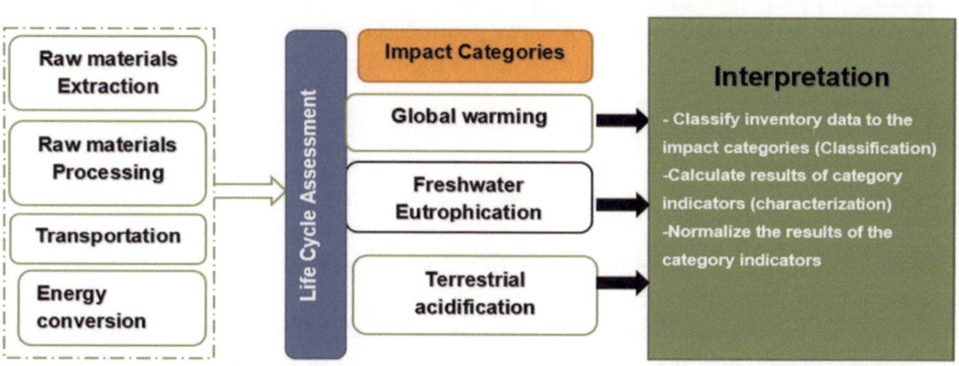

Figure 2.
System boundaries for electricity production using a natural gas conventional power process.

Materials/Input	Amount
Natural gas, high pressure	0.256 m^3
Gas powerplant 100 MW electrical	5.54e-10 unit
Water, completely softened	0.0598 kg
Water, decarbonised	1.99 kg
Residue from the cooling tower	−9.97e-06 kg
Water, cooling, unspecified natural origin	0.0589 m^3
Mode of transportation via pipeline	2.41e+12 ton/km.
Output	
Electricity	1 kWh
Emissions to air	
Carbon dioxide, fossil	0.533 kg
Methane	9.68e-06 kg
Nitrogen oxides	0.000366 kg
Sulfur dioxide	5.7e-06 kg
Emissions to water	
Water	0.06 m^3

Table 1.
List of input/output data of 1kWh of high voltage electricity from natural gas in a power plant [46].

database, which was integrated into the analysis software. This dataset represents the production of high-voltage electricity in a conventional steam boiler natural gas power plant without CHP (combined heat and power). A cradle-to-gate inventory involves all the processes/flows, raw materials and essential requirements needed to make electricity from the power plant available in the busbar and ready for distribution. LCA can be performed from either an attributional or a consequential perspective. An attributional LCI attempts to describe the environmentally relevant physical flows from and to a life cycle product system [43] and can be used to assess a product's environmental impact over time. Alternatively, consequential LCA describes how potential past or future decisions might have affected environmentally relevant physical flows [43–45]. This study used the attributional method resulting in the environmental implication of 1 kWh of electricity produced from the natural gas power plant. The inventory input/output data for 1kWh of high voltage electricity from natural gas in a power plant are included in **Table 1**. The analysis computation was based on the ROW (Rest of the World) data.

2.4 Life cycle impact assessment (LCIA)

The life cycle impact assessment (LCIA) is a process that converts the life cycle inventory (LCI) data into quantifiable environmental, social, and economic implications [47]. By converting the emissions and resource extractions contribute to environmental damage. Ratings, it aids in the interpretation of the evaluation [48]. It is a multi-issue technique for assessing potential environmental impact based on environmental resources (inputs and outputs) listed in the life cycle inventory. This is an attempt to link a product and its possible environmental impact. Flows are classified according to their environmental impact in the midpoint LCIA approach employed in this study. This method simplifies multiple flows by condensing them into a few common environmental impacts. This study, however, concentrates on three impact categories: global warming, eutrophication, and acidification. ReCiPe 2016 was used as the LCIA tool in this study [49, 50]. In SimaPro, GWP 100 factors are recommended as default in the Global Guidance for Life Cycle Impact Assessment Indicators and Methods (GLAM) [51]; the same is applied to GWP20 and GTP100 factors for sensitivity analysis. Those defaults are applicable to this study. More on ReCiPe can be found in the study of Ige et al. [50].

The LCIA (Life Cycle effect Assessment) process is founded upon the utilization of effect categories and characterization variables: classification, normalization, characterization, and valuation. Through classification, there is a grouping of the environmental impact measured to a sizeable recognized environmental impact category based on the availability of the process information. The characterization step's responsibility is to assess each environmental impact contribution [52]. This is done by multiplying each substance's amount by its characterization factor and finding the sum. The following equations from Huijbregts et al. [52]. study express the characterization formula types. Eq. (1) pertains to the variables that are generic in nature, whereas Eq. (2) pertains to the elements that are non-generic. The former factor is typically derived from characterization models and can be found in the literature as a database.

$$S_j = \sum_i Q_{j,i} m_i \tag{1}$$

where
S_j = impact category j indicator
m_i = size of the intervention of type i
$Q_{j,i}$ = characterization factor that links intervention i to impact category j

Eq. (2) denotes the potential variables of certain non-generic characterization aspects within the context of human health and the impact on the natural environment.

$$Q_{j,s,t} = \sum_l \frac{Effect(i,l,t)}{Emission(i,s)} = \sum_l \left(\frac{Fate(i,l,t)}{Emission(i,s)}\right) \cdot \left(\sum_l \frac{Exposure(i,l,t)}{Fate(i,l,t)}\right) \cdot \left(\sum_l \frac{Effect(i,l,t)}{Exposure(i,l,t)}\right) \tag{2}$$

where
subscript i = substance,
s = location of the emission,
l = related exposure area of the receptor
t = period during which the potential contribution to the impact is considered.

The normalization step is for comparison of the various impact across the impact classifications and inaccessible areas for prioritizing alternate product or resolving trade-offs between products [52, 53]. It is at this stage that the insignificant impact categories on the entire environmental impact result to the reduction of factors evaluated. According to Huijbregts et al. [52], normalization serves two primary objectives within the field of Life Cycle Impact Assessment (LCIA): firstly, it aims to situate the results of LCIA indicators within a wider framework, allowing for a more

Production unit	Processes considered
Extraction of raw materials (natural gas)	Natural gas, including the inputs and outputs
Processing of raw materials (natural gas)	Inputs and outputs required for the processing of natural gas
Transportation	The transportation of natural gas from the extraction location to the gate of the plant
Energy conversion	Every process involved to get electricity ready in the busbar of the power plant for distribution

Table 2.
The processes examined in each stage of production of natural gas burnt in a power plant.

comprehensive understanding of their implications. Secondly, it seeks to standardize the results by aligning them with common dimensions, facilitating meaningful comparisons and analyses. A reference value is employed to split the aggregate of the outcome for each indicator within a certain category.

$$N_k = {}^{S_k}\!/_{R_k} \tag{3}$$

where
k = impact category
N = normalized indicator
S = category indicator from the characterization phase
R = reference value.

The selection of the reference system is typically based on the overall indicator outcome for a specific country or region during a given year. In the context of a Life Cycle Assessment (LCA) study, the outcomes of normalization can facilitate the process of input grouping or weighting of effect categories, as well as provide a means to assess the relative significance of various impact categories [50, 53]. The inputs and outputs in the 1kWh of high voltage electricity from natural gas in power in **Table 1** are divided into four production stages: Extraction, processing, transportation, and energy conversion. The connection between each step and the impact categories under investigation is established through the utilization of the Life Cycle Assessment (LCA) methodology. SimaPro (version 9.2) software application was used for all calculations. **Table 2** shows all processes studied in each production stage. The mode of transportation in this study is a pipeline (2.41e+12 ton/km).

3. Results

The result represents data modeled after the ROW, including Austria, Belgium, Germany, Spain, France, the United Kingdom, Italy, Japan, Luxembourg, Netherlands and the United States of America. The following section explains the contribution of each unit process to the environmental impact and the major contributing processes throughout the life cycle. The environmental impact of each production process was studied in terms of Global Warming Potential (GWP), Freshwater Eutrophication Potential (FEP) and Terrestrial Acidification Potential (TAP). **Table 3** presents the total characterization results for the three selected impact categories. (The values

Impact category	Unit	Results
Global warming	kg CO_2 eq	6,05E-01
Freshwater eutrophication	kg P eq	8,28E-06
Terrestrial acidification	kg SO_2 eq	2,52E-04

Table 3.
Characterization results of the three focused environmental impacts.

provided in these tables are the functional unit of this current work, 1 kWh of electricity produced.

The GWP, FEP and TAP values for the impact results per functional unit (**Table 3**) are 6,05E-01 kg CO_2 eq, 8,28E-06 kg P eq and 2,52E-04 kg SO_2 eq, respectively. In this view, the impacts are mainly attributable to plant operation, with natural gas consumption and direct emissions to the atmosphere being the primary contributors. These results are consistent with the relevant scientific literature on electricity production from natural gas in power plants [54]. Further analysis was performed to determine each production stage's contribution to these Impact potentials in a natural gas power plant. These stages include Extraction, Processing, Transportation and Energy conversion.

Global warming potential: 79.5% of the emission is from the energy conversion stage; 14.3% is from the extraction stage, 3.4% is from the processing stage and 2.7% is from the transportation stage, as shown in **Figure 3**. From **Table 3**, the global warming impact category has the highest value; it qualifies as one with the highest impact and consequently presents the energy conversion process as the highest contributor to the environmental impact related to the production of electricity from natural gas power plants. Terrestrial Acidification: 79.3% is from the extraction stage; 11.9% is from the energy conversion stage; the rest is from the processing and transportation stages. Also, roughly 100% of the contribution to Freshwater Eutrophication is from emissions from the extraction stage for every production of 1 kW of electricity from the natural gas power plant.

The GWP is majorly experienced at the energy conversion stage, whereas the eutrophication and acidification potentials are primarily experienced during the extraction stage, as shown in **Figure 3**. In a gas power plant, the chemical energy

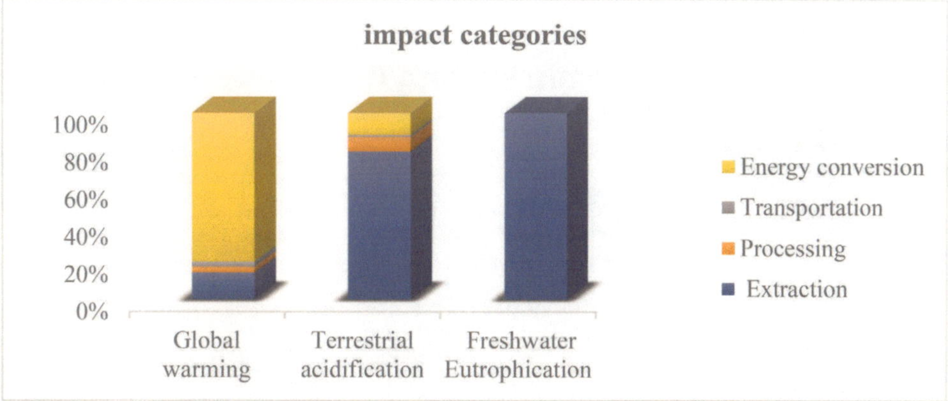

Figure 3.
Contribution of production stages to impact categories.

stored in the natural gas is converted into thermal, mechanical, and electrical energy. Thus, the maximum contribution to global warming observed due to climatic changes was recorded at the energy conversion stage (79.6%), which is understood to be the most demanding phase. Further analysis of the substance contributing to the GWP, FEP and TAP are discussed below.

3.1 Global warming potential (GWP)

The contributions of three GHGs, CO_2, N_2O, and CH_4, were considered when calculating the GWP, as shown in **Figure 4**. The global warming potential value of the process is 6,05E-01 kg CO_2 eq. as shown in **Table 3**.

The most significant contributor to GWP is CO_2 (94%), followed by CH_4 (5%) and N_2O (1%) emissions as shown in **Figure 4**. CO_2 is emitted in the most considerable quantity of air emission due to the fuel burning caused by global warming. The percentage value of CH_4 is attributed to the fugitive emissions from natural gas production. Most CH_4 results are from natural gas losses during raw material extraction and transportation. The energy conversion stage is the largest source of GHG emissions due to gas burning, accounting for 79.5% of total GWP.

3.2 Freshwater eutrophication potential (FEP)

The value of the Freshwater Eutrophication impact is 8,28E-06 kg P eq./kWh, as shown in **Table 3**. Phosphate (PO_4^{3-}) emitted 99.7% into the waterbody and Phosphorus (P water) is 0.3%, as shown in **Figure 5**. (PO_4^{3-}) is the major substance that contributed to the FEP.

Phosphorus (P waster) can directly regulate algae growth in aquatic ecosystems as vital nutrition, which has been recognized as a limiting factor for eutrophication [55]. Although PO_4^{3-} is soluble reactive phosphorus, algae can preferentially absorb it. Phosphorus (P) pollution can *trigger severe marine eutrophication*, leading to harmful algal blooms and seawater deterioration. Too much phosphorus can cause increased growth of algae and large aquatic plants, which can result in decreased levels of dissolved oxygen– a process called eutrophication.

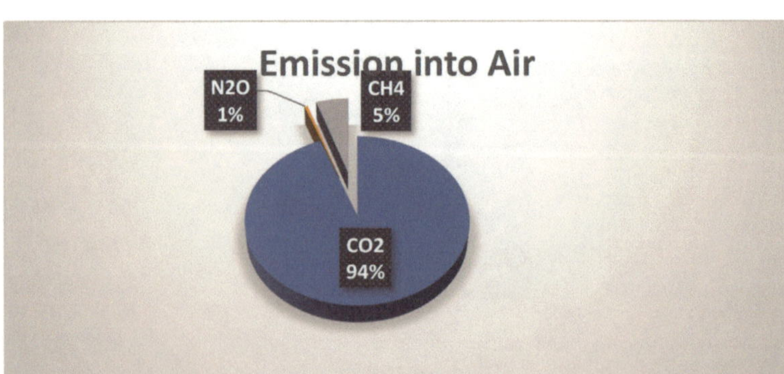

Figure 4.
Contribution results for global warming potential.

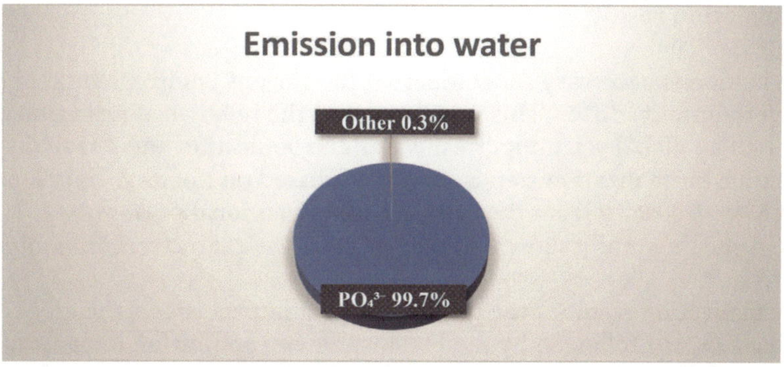

Figure 5.
Contribution results for freshwater eutrophication.

3.3 Terrestrial acidification potential (TAP)

The terrestrial acidification value of the 1 kWh electricity production from the natural gas conventional power plant is 2,52E-04 kg SO_2 eq./kWh, as shown in **Table 3**. NO_x emitted 72% of the emission, followed by SO_2 with 27%, while NH_3 contributed 1%, as shown in **Figure 6**. This situation involves the extraction of natural gas as a raw resource. Burning fossil fuels containing sulfur, such as natural gas, always produce SO_2.

Sulfur dioxide (SO_2) has been found to be a significant contributor to respiratory ailments in human populations, while also playing a substantial role in the formation of acid rain. The planet Earth. The process of anthropogenic ozone generation begins with the release of nitrogen oxides (NO_x) and/or non-methane volatile organic compounds (NMVOCs) into the atmosphere. Through subsequent chemical reactions, the ozone layer is generated. The elevated levels of ozone production in the Earth's atmosphere have significant implications for both human health and the overall ecology. The impact of this phenomenon is evident in the occurrence of health complications and, in some cases, the extinction of certain species.

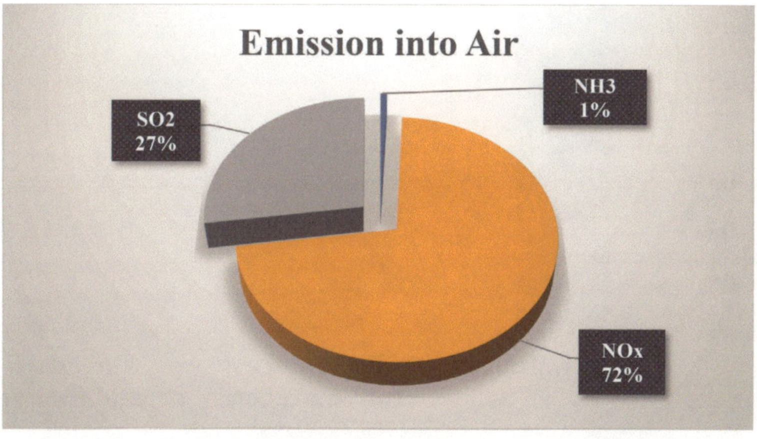

Figure 6.
Contribution results for terrestrial acidification.

3.4 Normalization result

Normalization is necessary for comparing the various environmental impact categories since their units differ. This step presented the relative contribution of each impact caused by global warming, freshwater eutrophication and terrestrial acidification. A normalization step is conducted based on the total emissions to produce 1 kWh of electricity from the natural gas conventional power plant. The normalization results naturally show a similar trend as the characterization impact results (**Figure 7**).

The normalization result of the global warming impact shows the highest value (7,57E-05 kg CO_2 eq) followed by the Freshwater eutrophication impact value of 127509E-05 kg P eq./kWh and Terrestrial acidification impact with a value of 615531E-06 kg SO_2 eq./kWh. The most harmful impact category in the normalization result is global warming due to the burning natural gas in the power plant.

4. Conclusion

This study examined the environmental impact of electricity production from the natural gas power plant. In this study, we used a cradle-to-gate method as a system boundary. Cradle-to-gate includes raw material extraction (natural gas), raw material processing (natural gas), transportation and energy conversion stages. The results of this study helped identify the environmental sustainability of 1 kWh of electricity production from a natural gas conventional power plant. The results show that the GWP is calculated at 6,05E-01 kg CO_2 eq./kWh. According to this study, CO_2 accounts for 94%, CH_4 5%, and N_2O 1% of all air emissions are the three main sources of GWP. Electricity production from the natural gas conventional power plant is estimated at 8,28E-06 kg P eq./kWh for FEP. The TAP of 2,52E-04 kg SO_2 eq./kWh. According to the analysis, the environmental impact assessment showed a good

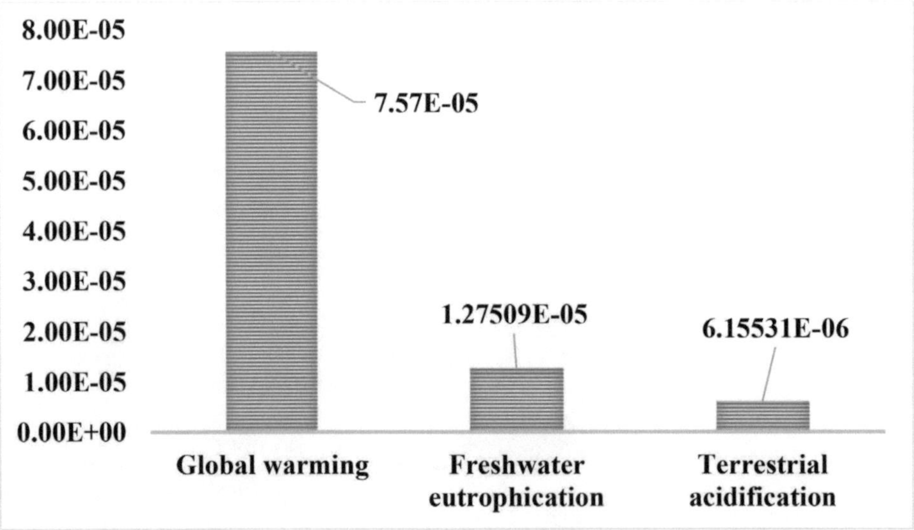

Figure 7.
Normalized results of each impact category.

environmental performance compared with other literature [16, 56]. The result showed that the environmental impact hotspot was the raw material processing (natural gas) stage.

Regarding production stages, the energy conversion stage (79.57%) is the main hotspot of the GWP. The raw material extraction, processing and transportation are often insignificant. At the normalization step, global warming impact with the values of 7,57E-05 kg CO_2 eq is the most harmful environmental impact category. This work discussed the environmental implication of 1 kWh of electricity production from a natural gas conventional power plant. The LCA methodology used on natural gas for electricity production shows that the results depend on system boundaries, the data source and the technologies used. Further improvement on the environmental performance would require careful component selection for the right technology to allow mitigation. LCA shows that natural gas conventional power plant for electricity production is more environmentally friendly than other fossil fuels. The result of this study can be used as a guide for stakeholders involved in the environmental implication of plants and policymakers are bound to understand better how the electricity production from the natural gas conventional power plant is allocated. Also, the results of this study are relevant since natural gas is being promoted globally as a fuel source for electricity-producing plants.

Future studies will include a comparison of these results with other energy technologies. Comparison of different capacity sizes of natural gas power plants should consist of other indicators apart from Global Warming Potential, Freshwater Eutrophication Potential and Terrestrial Acidification Potential for impact assessment.

Author details

Oludolapo Akanni Olanrewaju[1*], Oluwafemi Ezekiel Ige[1], Busola Dorcas Akintayo[1] and Ahad Ali[2]

1 Durban University of Technology, Durban, South Africa

2 Lawrence Technological University, Michigan, USA

*Address all correspondence to: oludolapoo@dut.ac.za

References

[1] Ahmadi GR, Toghraie D. Energy and exergy analysis of Montazeri steam power plant in Iran. Renewable and Sustainable Energy Reviews. 2016;**56**:454-463

[2] C. by fuel type-Exajoules and C. D. Emissions, bp Statistical Review of World Energy. London, UK: BP PLC; 2020. [Online]. Available from: http://www.bp.com/statisticalreview

[3] IEA. Electricity Demand by Sector and Scenario, 2018–2040. Paris: IEA; 2020. Available from: https://www.iea.org/data-and-statistics/charts/electricity-demand-by-sector-and-scenario-2018-2040 [Accessed Feburary 2 2023]

[4] Rogelj J, et al. Mitigation pathways compatible with 1.5 C in the context of sustainable development. Intergovernmental Panel on Climate Change. Geneva: Global warming of 1.5°C (IPCC Special Report); 2018. pp. 93-174

[5] Watch C. Global Historical Emissions. USA: World Resources Institute; 2021. Available from: https://www.climatewatchdata.org/ghg-emissions

[6] Pedersen JST et al. An assessment of the performance of scenarios against historical global emissions for IPCC reports. Global Environmental Change. 2021;**66**:102199

[7] Gould T, McGlade C. The environmental case for natural gas. France: International Energy Agency (IEA); 2017. [Online] Available from: https://www.iea.org/commentaries/the-environmental-case-for-natural-gas [Accessed: October 1, 2023]

[8] Riva A, D'Angelosante S, Trebeschi C. Natural gas and the environmental results of life cycle assessment. Energy. 2006;**31**(1):138-148

[9] Swift B. Cleaner Power: The Benefits and Costs of Moving from Coal Generation to Modern Technologies. Washington, DC: Environmental Law Institute; 2001

[10] Speight J. Recovery, storage, and transportation. Natural Gas. 2019;**2019**: 149-186

[11] Boretti A. Advances in diesel-LNG internal combustion engines. Applied Sciences. 2020;**10**(4):1296

[12] Shasby BM. Alternative Fuels: Incompletely Addressing the Problems of the Automobile. Alexandria, Virginia: Masters Urban and Regional Planning, Virginia Polytechnic Institute and State University; 2004

[13] Khan MI, Yasmin T, Shakoor A. Technical overview of compressed natural gas (CNG) as a transportation fuel. Renewable and Sustainable Energy Reviews. 2015;**51**:785-797

[14] Wu X et al. Comparative life cycle assessment and economic analysis of typical flue-gas cleaning processes of coal-fired power plants in China. Journal of Cleaner Production. 2017;**142**: 3236-3242

[15] Alizadeh S, Avami A. Development of a framework for the sustainability evaluation of renewable and fossil fuel power plants using integrated LCA-emergy analysis: A case study in Iran. Renewable Energy. 2021;**179**: 1548-1564

[16] Roinioti A, Koroneos C. Integrated life cycle sustainability assessment of the Greek interconnected electricity system. Sustainable Energy

Technologies and Assessments. 2019;**32**: 29-46

[17] Kabayo J, Marques P, Garcia R, Freire F. Life-cycle sustainability assessment of key electricity generation systems in Portugal. Energy. 2019;**176**: 131-142

[18] Mahmud MP, Farjana SH. Comparative life cycle environmental impact assessment of renewable electricity generation systems: A practical approach towards Europe, North America and Oceania. Renewable Energy. 2022;**193**:1106-1120

[19] Mahmud MP, Huda N, Farjana SH, Lang C. Environmental impacts of solar-photovoltaic and solar-thermal systems with life-cycle assessment. Energies. 2018;**11**(9):2346

[20] de Haes HAU, Heijungs R. Life-cycle assessment for energy analysis and management. Applied Energy. 2007;**84** (7–8):817-827

[21] Bhat I, Prakash R. LCA of renewable energy for electricity generation systems—A review. Renewable and Sustainable Energy Reviews. 2009;**13**(5):1067-1073

[22] Muench S, Guenther E. A systematic review of bioenergy life cycle assessments. Applied Energy. 2013;**112**: 257-273

[23] Nie Z, Korre A, Durucan S. Life cycle modelling and comparative assessment of the environmental impacts of oxy-fuel and post-combustion CO2 capture, transport and injection processes. Energy Procedia. 2011;**4**:2510-2517

[24] Eliasson B, Riemer P, Wokaun A. Greenhouse Gas Control Technologies. Oxford, United Kingdom: Elsevier Science Ltd.; 1999. p. 1223

[25] Barros MV, Salvador R, Piekarski CM, de Francisco AC, Freire FMCS. Life cycle assessment of electricity generation: A review of the characteristics of existing literature. The International Journal of Life Cycle Assessment. 2020;**25**:36-54

[26] Dzikuć M, Piwowar A. Ecological and economic aspects of electric energy production using the biomass co-firing method: The case of Poland. Renewable and Sustainable Energy Reviews. 2016; **55**:856-862

[27] Garcia R, Marques P, Freire F. Life-cycle assessment of electricity in Portugal. Applied Energy. 2014;**134**:563-572

[28] Rodríguez MR, Cespón MF, De Ruyck J, Guevara VO, Verma V. Life cycle modeling of energy matrix scenarios, Belgian power and partial heat mixes as case study. Applied Energy. 2013;**107**:329-337

[29] Messagie M et al. The hourly life cycle carbon footprint of electricity generation in Belgium, bringing a temporal resolution in life cycle assessment. Applied Energy. 2014;**134**: 469-476

[30] Turconi R, Tonini D, Nielsen CF, Simonsen CG, Astrup T. Environmental impacts of future low-carbon electricity systems: Detailed life cycle assessment of a Danish case study. Applied Energy. 2014;**132**:66-73

[31] Laurent A, Espinosa N. Environmental impacts of electricity generation at global, regional and national scales in 1980–2011: What can we learn for future energy planning? Energy & Environmental Science. 2015; **8**(3):689-701

[32] Akber MZ, Thaheem MJ, Arshad H. Life cycle sustainability assessment of

electricity generation in Pakistan: Policy regime for a sustainable energy mix. Energy Policy. 2017;**111**:111-126

[33] Barros MV, Piekarski CM, De Francisco AC. Carbon footprint of electricity generation in Brazil: An analysis of the 2016–2026 period. Energies. 2018;**11**(6):1412

[34] Silva DAL, Delai I, Montes MLD, Ometto AR. Life cycle assessment of the sugarcane bagasse electricity generation in Brazil. Renewable and Sustainable Energy Reviews. 2014;**32**:532-547

[35] Ige OE, Olanrewaju OA, Duffy KJ, Collins OC. A review of the effectiveness of life cycle assessment for gauging environmental impacts from cement production. Journal of Cleaner Production. 2021;**2021**:129213. DOI: 10.1016/j.jclepro.2021.129213

[36] ISO. 14040: International organization for standardization. Environmental Management: Life Cycle Assessment; Principles and Framework. International Organization for Standardization. ISO. 2006. Available from: https://www.iso.org/obp/ui/#iso:std:iso:14040:ed-2:v1:en [Accessed: June 12, 2022]

[37] Jiménez-González C, Kim S, Overcash MR. Methodology for developing gate-to-gate life cycle inventory information. The International Journal of Life Cycle Assessment. 2000;**5**:153-159

[38] Filimonau V. Life Cycle Assessment (LCA) and Life Cycle Analysis in Tourism. Berlin/Heidelberg, Germany: Springer; 2016

[39] Pré Consultants B. SimaPro. 2016. Available from: https://network.simapro.com/esuservices [Accessed: August 13, 2022]

[40] Goedkoop M, De Schryver A, Oele M, Durksz S, de Roest D. Introduction to LCA with SimaPro 7. The Netherlands: PRé Consultants; 2008

[41] Olagunju BD, Olanrewaju OA. Life cycle assessment of ordinary Portland cement (OPC) using both problem oriented (midpoint) approach and damage oriented approach (endpoint). In: Product Life Cycle-Opportunities for Digital and Sustainable Transformation. London, UK: IntechOpen; 2021

[42] Finkbeiner M, Inaba A, Tan R, Christiansen K, Klüppel H-J. The new international standards for life cycle assessment: ISO 14040 and ISO 14044. The International Journal of Life Cycle Assessment. 2006;**11**:80-85

[43] Ekvall T, Tillman A-M, Molander S. Normative ethics and methodology for life cycle assessment. Journal of Cleaner Production. 2005;**13**(13–14):1225-1234

[44] Ekvall T, Weidema BP. System boundaries and input data in consequential life cycle inventory analysis. The International Journal of Life Cycle Assessment. 2004;**9**(3):161-171

[45] Finnveden G et al. Recent developments in life cycle assessment. Journal of Environmental Management. 2009;**91**(1):1-21

[46] Treyer K. Electricity Production, Natural Gas, Conventional Power Plant. Zürich, Switzerland: Ecoinvent Association; 2019. Available from: https://v36.ecoquery.ecoinvent.org/Details/PDF/5EEE12FA-FB06-4BB6-97E3-E8FDDBCF6C96/290C1F85-4CC4-4FA1-B0C8-2CB7F4276DCE [Accessed: September 16, 2022]

[47] Goglio P et al. A comparison of methods to quantify greenhouse gas

emissions of cropping systems in LCA. Journal of Cleaner Production. 2018;**172**: 4010-4017

[48] Hauschild MZ, Huijbregts MA. Introducing Life Cycle Impact Assessment. Netherlands: Springer; 2015

[49] Lundie S, Peters GM. Life cycle assessment of food waste management options. Journal of Cleaner Production. 2005;**13**(3):275-286

[50] Ige OE, Olanrewaju OA, Duffy KJ, Collins OC. Environmental impact analysis of Portland cement (CEM1) using the midpoint method. Energies. 2022;**15**(7):2708 Available from: https://www.mdpi.com/1996-1073/15/7/2708

[51] Frischknecht R et al. Global guidance on environmental life cycle impact assessment indicators: Progress and case study. The International Journal of Life Cycle Assessment. 2016;**21**:429-442

[52] Huijbregts MAJ et al. Priority assessment of toxic substances in life cycle assessment. Part I: Calculation of toxicity potentials for 181 substances with the nested multi-media fate, exposure and effects model USES–LCA. Chemosphere. 2000;**41**(4):541-573. DOI: 10.1016/S0045-6535(00)00030-8

[53] Finnveden G et al. Normalization, grouping and weighting in life cycle impact assessment. In: Life Cycle Impact Assessment: Striving Towards Best Practice. Pensacola, FL: Society of Environmental Toxicology and Chemistry (SETAC); 2002

[54] Treyer K, Bauer C. Life cycle inventories of electricity generation and power supply in version 3 of the ecoinvent database—Part I: Electricity generation. The International Journal of Life Cycle Assessment. 2016;**21**(9): 1236-1254

[55] Clarke A et al. Long-term trends in eutrophication and nutrients in the coastal zone. Limnology and Oceanography. 2006;**51**(1):385-397

[56] Turconi R, Boldrin A, Astrup T. Life cycle assessment (LCA) of electricity generation technologies: Overview, comparability and limitations. Renewable and Sustainable Energy Reviews. 2013;**28**:555-565